THE ECONOMICS OF FORESTRY
Modern Theory and Practice

THE ECONOMICS OF FORESTRY
Modern Theory and Practice

Erhun Kula
Department of Economics, University of Ulster
Northern Ireland, UK

CROOM HELM
London & Sydney

TIMBER PRESS
Portland, Oregon

First published in 1988 by Croom Helm
11 New Fetter Lane, London EC4P 4EE
Croom Helm Australia, 44-50 Waterloo Road,
North Ryde, 2113, New South Wales

Printed in Great Britain by
St Edmundsbury Press Limited
Bury St Edmunds, Suffolk.

ISBN 0 7099 3145 X

British Library Cataloguing in Publication Data

Kula, Erhun
The economics of forestry : modern theory and practice.
1. Forests and forestry—Economic aspects
I. Title
338.1′749   SD393

ISBN 0–7099–3145–X

First Published in the USA 1988 by
Timber Press
9999 S.W. Wilshire
Portland, OR 97225
USA

ISBN 0-88192-123-8

Library of Congress Cataloging in Publication Data

Kula, Erhun.
The economics of forestry : modern theory and practice / Erhun Kula.
p.   cm.
Bibliography: p.
Includes index.
ISBN 0-88192-123-8
1. Forests and forestry—Economic aspects.   I. Title.
SD393.K82 1988
338.1′349—dc19

Contents

To my son

About the author

Erhun Kula is a Turkish-born economist who did his first degree in Istanbul. After completing his M.Sc. at the University of Wales and his Ph.D. at the University of Leicester, he taught for a couple of years at the Centre for Development Studies, University College, Swansea. Currently he is a Senior Lecturer in Economics at the University of Ulster, Northern Ireland. He has published in many learned journals of economics and environmental science in the fields of cost-benefit analysis, forestry, fishing, agriculture and famine anticipation.

Preface

The economics of forestry has always fascinated me as one of the most brain-taxing cases in economics. As an investment forestry is different from many other projects as it has unusually long gestation periods. For example, in the United Kingdom it takes over 40 years to grow coniferous and over 100 years for deciduous timber.

These long gestation periods make it very clear how important are the magnitude of the discount rate and the method of discounting in the evaluation of investment projects. Any errors in these will misguide investors in forestry one way or the other.

In addition, forestry redistributes income between generations. Its long gestation periods make it obvious that more than one generation will be involved in any venture. When we plant trees we know that the bulk of the benefits will be captured by future generations. Conversely, when we fell trees we reap the benefit of projects which were established in the past, mostly by generations who are long gone. So far most economists have devoted their time and energy to analysing income distribution in an intragenerational context, and this is a very sensitive and controversial issue. After all, most revolutions have taken place because of the uneven income distribution which was oppressive for the majority. Forestry helps us to study the case from the viewpoint of different generations.

Forestry necessitates estimating timber prices a long time ahead. No other investment project makes such burdensome demands on economists. In this book I have a great deal to say about the first two aspects of forestry. As regards the third case, i.e. price projection for the distant future, I have declined to enter into deep discussion on this matter in view of the time scale involved: could anybody be expected to make an accurate guess as to the price of, say, veneer oak 120 years from now?

The bulk of the ideas which are presented in this book stem from my work and resulting publications in journals such as *Environment and Planning A*, *Journal of Agricultural Economics*, *Journal of Development Studies*, *Quarterly Journal of Economics*, *Forestry*, *Resources Policy* and *Project Appraisal*. I am grateful to these journals for granting me permission to reproduce some parts of my material, i.e. tables, figures, etc. I have also used my teaching notes for the course 'Economics of Exhaustible Resources', which I have been teaching at the University of Ulster during the last five years.

I am well aware of the fact that some of my ideas, particularly on discounting, will not be easily digested by some economists. I have spent the last ten years of my life studying the case of discounting, especially in public sector economics. I started out on this mission when I was an enthusiastic and idealistic research student. Many of my fellow economists tried to dismiss this work simply because I was trying to say something radically new and hoping to correct past errors which were affecting not only our own lives, but also those of future generations.

The book should be of interest to economists, foresters, geographers and environmental scientists as the material presented is useful in all these subject areas. Chapter 1 explains, first, how in the British Isles forests were almost completely destroyed in the course of history, a destruction which was repeated to varying degrees, in many other Western countries. When the British government decided to reverse the trend, it came up against a fundamental economic difficulty, discounting. The emphasis is placed upon a number of events by which the principles of discounting were badly and fatally twisted to justify afforestation projects. The chapter brings it home to economists that the general economic principles designed to analyse public sector projects did not work out for forestry. Something was wrong somewhere. It also reminds foresters that their projects cannot escape economic scrutiny by just becoming a special case.

Chapter 2 explains, briefly, the history and the current structure of forestry in a number of Western countries and the EEC. It is written for economists. Highly informed foresters are likely to be knowledgeable about these subjects and therefore may skip this chapter.

Chapter 3 carries out a cost–benefit analysis of forestry as practised in Britain and Ireland. It uses the widely practised ordinary discounting principle, which is inaccurate.

Chapter 4 explains what is wrong with the ordinary discounting method and provides an alternative, modified discounting, not only for forestry but for all public sector investments. The new method is used in the economic evaluation of public sector forestry with dramatic results. It is also emphasised that the modified discounting method would change the project mix in the public sector portfolio towards the inclusion of more long-term projects.

Chapter 5 is a study of private sector afforestation projects which, thanks to a number of government incentives, is a booming sector in Britain. The planting grants and tax concessions given to private investors have made forestry an attractive proposition in recent years. In this chapter financial rates of return are calculated for investors in two different locations, poor and moderate. Chapter 5 also reports on a recent case study in which the responsiveness of private foresters to changes in a number of variables, such as government grants, timber prices, agricultural subsidies and land prices, are measured. Here readers are warned that in calculations modern econometric technology is used and those who are not familiar with it should avoid the mechanics of the models used and concentrate on the final results and their implications for private sector forestry.

The last chapter deals with the old and rather difficult problem of finding the optimal rotation period for felling a tree or a stand of trees. Here every effort has been made to avoid unnecessary complications, and a number of worked-out examples are provided. Chapter 6 ends with the conclusion that in public sector forestry the modified discounting method will make a substantial impact on finding the optimum rotation.

I should like to express my thanks to a number of individuals who helped me along the way. David W. Pearce was one of the economists who had sympathy with my ideas, which helped at a time when I needed it most. Roger Cooper made many helpful comments on the early manuscript of this book. Colin Price was helpful on occasion in explaining to me details of forestry in the United Kingdom. Of course, his major contribution is in the launching of a lively debate which is still continuing in the literature. My work with Donal McKillop, presented in Chapter 5, was also a great help in providing further insights into forestry economics. I am also grateful to him for reading parts of the manuscript. Many staff members of the Forest Service, Department of Agriculture, Northern Ireland, and in particular Dr R. McKenzie, helped me in compiling research material. Last but not least I am immensely grateful to my wife on two counts: for typing most of the manuscript, and for giving support when the pressure of work was great.

Erhun Kula
Coleraine, Northern Ireland, 1987

1

Forestry Policy: An Historic Overview

Forestry policy in almost every country has been shaped by historic events. Therefore, a study of forestry policy would be incomplete without giving a proper account of the historic background. In the past there were three main events which led to the contraction of forestry throughout the industrialised world. First, the growth of the population increased the pressure on agricultural land to grow more food, which resulted in forest clearances. The decline of forests in many regions became inevitable to sustain the growing agricultural sector. Secondly, the use of wood as fuel and construction material took its toll on forests, and such exploitation was most reckless in the early industrialised countries. Thirdly, wars had a devastating effect on forests, especially in Europe. The scars of these events have been deepest in the British Isles.

Unfortunately, history seems to be repeating itself in the Third World. In their haste to catch up with the West and to feed their rapidly growing population, many developing countries are putting an overwhelming emphasis on agricultural and industrial expansion, without due regard to forestry. Although wood is a renewable resource, the regeneration of the forestry sector is a painfully slow process, a case which will be explained in this chapter with reference to the British experience. It is essential that the Third World countries should not repeat the mistakes that were made by some industrialised countries in the past when they recklessly depleted their timber deposits, from which deforestation they have not yet recovered.

Towards the end of the Middle Ages, the British Isles were clothed with thick forests. Unfortunately, today they are among the most treeless areas in Europe. The history of Britain's forests can be divided into three distinct periods: early historic times and the Middle Ages; from the end of the Middle Ages to the 20th century; and the 20th century.

EARLY HISTORIC TIMES AND THE MIDDLE AGES

Geologists recognise four main glaciations in the British Isles, the last of which was centred near Ireland. During the last glaciation and for a long time afterwards there were no forests in these islands. Trees started growing well after the recession of the ice around 8000–7500 BC. Birch, pine and willow were the earliest species. During the period between 7500 and 5500 BC the climate became warmer. The mean summer temperature was probably 2–3°C higher than it is today. The warm climate encouraged the spread of other species such as hazel, oak and elm, giving way to dense forests for the first time. From 5500 to 2500 BC the climate continued warm but became wetter, and as a result some existing species such as oak and elm established dominance in many areas. The wet climate also favoured some new species such as elder and holly which took hold especially in the west and the north. Hazel scrub and sphagnum bog developed in some exposed western areas at the expense of previously established vegetation.

Man began to put pressure on forests for the first time in the Bronze Age. During this period clearance took place mainly for agricultural purposes. After 550 BC the climate became cooler. Peat and bog spread, destroying pine and birch forests especially in the west. Man aided this destruction by clearing trees from the plains, a process which was to continue for hundreds of years. However, this was not extensive and can be described as merely ‘nibbling around the edges’ of great forests.

During the Middle Ages destruction gathered pace. By and large forests in the British Isles were treated as common property resources. There were some privately owned forests, but private ownership did not prevent common folk from exploiting them for timber, nuts and fruit, firewood and hunting, and for running swine into them. In most cases forests were unappropriated or owned commonly by the inhabitants of a region. The common ownership inevitably led to reckless exploitation. From the time of William the Conqueror highly effective laws were passed to encourage conservation, but the purpose of these laws was to protect wild beasts in forests for hunting purposes. The timber value of forests was a somewhat secondary consideration. At the time the British Isles were still covered with thick forests and the supply of wood appeared to be inexhaustible. After the signing of the Magna Carta in the thirteenth century the early rules on forest utilisation were relaxed and as a result timber stocks began to decline along with the wildlife.

FROM THE END OF THE MIDDLE AGES TO THE TWENTIETH CENTURY

This is the saddest period in the history of forestry as many forests in the British Isles were destroyed almost completely. There were many reasons for this destruction. First, most of the forests located on fertile soils had to be cleared to make room for agricultural expansion to support the growing population. Secondly, clearance was necessary for the purpose of providing fuel for the iron smelting industry which was booming in the sixteenth and seventeenth centuries. This was so extensive that, at the end of the seventeeth century, turf firing had to be introduced in many areas in place of firing with wood. Around this period a traveller to Scotland wrote that 'a tree in here is rare as a horse in Venice' (Thompson, 1971). Third, the rise of Britain as a major industrial and naval power took its toll on forests. The industrialisation process was based entirely on a single fuel, coal. Timber has always been important as a source of pit props in the coal-mining industry. An expansion of this industry meant a decline in forests. Also timber was widely used in the shipbuilding industry, and rapid reductions of forests created anxiety amongst naval experts. But this did not last long. New discoveries overseas brought home the 'good news' that the world was full of countries overflowing with timber resources. Many thought that, if domestic stocks were to disappear, timber could always be imported from the colonies, or alternatively ships could be built there. However, not everyone shared this view. A commission proposed in 1786 that 100 000 acres of oak should be planted for the Navy, and this was carried out mostly within the Crown Estates. The building of the first iron ship for the Royal Navy in 1860 signalled the end of the shipbuilding industry's insatiable demand for oak. Fourthly, destruction also took place for military purposes, especially in Ireland. During the colonial struggle settlers cleared the strategic locations of trees because they provided cover for local resistance groups. Such clearance was particularly extensive in Ulster.

Towards the end of the eighteenth century a revival of forests took place in many parts of the British Isles. At the time the British aristocracy realised that woodlands were a desirable source of wealth and amenity which led to extensive planting for commercial as well as ornamental reasons. In 1775 the London Society of Arts was offering prizes to tree planters. In 1785 the Royal Dublin Society decided to pay out grants, £3 per acre, to private foresters. Forests in private estates were retained generation after generation for reasons of sentiment and prestige. Selling trees was usually an indication of the decline of the family.

The decline of private forests began around the middle of the nineteenth century, a time when many landlords, for various reasons, were in financial difficulties. Travelling mills moved into many estates to clear the land of trees. At the turn of this century the destruction was complete and the islands that once were rich in forests became more or less treeless. Tree coverage of the total land surface was under 2 per cent in places such as Ireland and Scotland.

THE TWENTIETH CENTURY

The year 1903 was an important one for forestry in the British Isles as a forestry branch of the Department of Agriculture was established in Ireland for the purpose of training young men as practical foresters. The Department also bought land in various areas in order to establish forestry centres with a view to afforestation. Avondale in County Wicklow was the first forestry centre where a training school was established.

Progress was slow to start with but it gathered pace in later years and the Department set up many other similar centres elsewhere in Ireland. Unfortunately, the First World War put a stop to the expansion of forestry projects. During the war Britain suffered a severe shortage of timber. The German submarine campaign reduced imports to almost nothing and domestic stocks were nearly exhausted. At the time timber was a strategic commodity used extensively for pit props — the country's vital source of energy being coal. Many found this situation intolerable as problems co-ordinating timber supplies increased. Forest exploitation was carried out partly by the established trade and partly by the hastily created Timber Supplies Department. There were great logistical problems in shifting timber from the remote areas to places where it was needed. Transportation was carried out by horses and trains and in many areas there were severe shortages of horses as well as manpower. The administrative posts of the Timber Supplies Department were quickly filled by university-trained foresters, and many prisoners of war were sent to do the manual work. After the war the Prime Minister, Lloyd George, admitted that Britain was nearer to losing the war from lack of timber than from lack of food.

In 1917 the Acland Committee was appointed to examine the implications of insufficient supplies of home-grown timber. It recommended establishing some kind of forestry authority to remedy the timber deficiency. The committee also recommended that in the long term towards the end of the twentieth century, 1770000 acres should be planted. On the basis of this recommendation, in 1919 the Forestry Commission was established. Some of the members of the Timber Supplies Department became the backbone of this Commission. As a first step it was hoped that, within the first decade, the Commission would create 200000 acres of new plantations as well as restoring 50000 acres of pre-existing forests. It was also agreed to encourage private planters by giving them state aid. Unfortunately, this early target figure of 250000 acres was not achieved because a persuasive anti-forestry lobby convinced the government that state forestry was not a profitable venture. As a result the government imposed an expenditure cut during the 1922–24 period to slow down the afforestation programme.

Then came the Second World War which reminded the government of the reasons why the Forestry Commission was established in the first place. Bulky timber imports by sea proved to be an easy target for the German Navy, and the trees planted in the 1920s were not ready for felling. Nevertheless, the production of home-grown timber was increased to many times the level during peacetime and this took its toll mainly on the mature and remote forests. Although the availability of mechanical equipment speeded up the haulage operation, there were many logistical problems. Established sawmillers of home-grown timber increased the capacity of their mills, and many new sawmills were erected in various parts of the country. At the beginning of the war, maximum prices were fixed for round timber and these prices eventually became standard over the whole of the United Kingdom. These prices favoured particularly those regions where prices had previously been low. Estates in favourable locations with good timber stocks scarcely benefited from this practice, but timber which was too distant or too poor to attract customers before the war became saleable on the new terms. It became possible almost everywhere to sell even the most inferior quality timber and the war allowed a general cleaning-up of the woods.

The demand for softwoods was far greater than for hardwoods, so the broad-leaved trees of the south suffered less severely than the coniferous forests of the north. The demand for wood was so great that the shortage continued for a long time after the war. Eventually the controlled prices failed to keep pace with the rise in wages or with the falling value of money. At the end of the war some of the plantations were sold at a price which was not high enough to meet the cost of re-establishment. In January 1947, prices of standing timber were raised by 25 per cent. Price control was abolished in December 1949 and free market forces began to operate, doubling many prices immediately. However, control was kept over the volume of felling, and licences were required to carry out this operation. In 1951–52 the quota was fixed at 1.5 million hectares of softwoods. This was a substantial reduction on wartime felling which was not very different from the annual felling between the wars. These licences were given to the owners of the timber and not to the merchants who bought it, and there was keen competition to buy licensed trees. The largest customers were the National Coal Board, which needed large quantities of wood for pit props; the General Post Office, which wanted timber for telegraph poles; and the construction industry, which was booming following the war.

Having had two bitter experiences, after the war further and quite extensive plantations were created in many parts of Britain and forests in general have enjoyed a period of growth under the protective policies of post-war governments. Today the Forestry Commission is the largest landowner in Britain and it manages well over 1.5 million hectares of land. The Commission also gives aid to private foresters, and at present there are over 1.5 million hectares of privately owned plantations. The duties of the Forestry Commission are: buying land, planting trees, harvesting and selling timber, establishing forest recreation centres for the benefit of the community, aiding private foresters by giving them advice and cash grants, doing research, and employing and training manpower to carry out all of these activities.

When the government of Northern Ireland was formed in 1921 it assumed responsibility for forestry in the province, and the new Department of Agriculture became the forestry authority. Today the Department of Agriculture's Forestry Division is in charge. This is a separate body from the Forestry Commission but has similar powers and duties to it. In the Republic of Ireland the relevant government authority is the Forestry and Wildlife Service. The Irish Government's aim is to expand forestry projects as fast as possible, hopefully with the help of the EEC, since Ireland, despite her very favourable climatic and soil conditions, has, after Iceland, the fewest trees in Europe.

The rigour of forestry policy in the United Kingdom can be adjusted in line with changing domestic and world conditions regarding timber demand and supply. Economic circumstances may lead to future governments reconsidering their investment in forestry as a result of major shifts in the domestic market. As for the global needs it is worth mentioning that currently the world's natural forests are under great pressure and future supplies are bound to rely more on plantations. Therefore, in the future a timber famine is more likely than a glut. For these reasons the government intends to carry out five-yearly reviews to judge the progress of forestry programmes in the light of changing circumstances.

At present, forestry policy in the United Kingdom is governed by three factors. First, there is the compelling need, demonstrated by two world wars, to reverse the process of deforestation and create domestic resources of timber. It is now well known that dairy and beef sectors are over-expanded in the United Kingdom and some other Northern countries in the EEC, creating surplus output, a source of embarrassment to the EEC governments. On the other hand the United Kingdom and many other EEC countries are continuing to import large quantities of timber from outside. In effect, at present the EEC is only 50 per cent self-sufficient in wood. The figure for the

UK is about 10 per cent. The import bill for forest products in the United Kingdom amounted to about £4.5 billion in 1986, making it the second largest import item after food. In many people's minds over-production of dairy and beef and under-production of timber are powerful signs of misallocation of land between the sectors in the country. Secondly, there is a need to provide productive work in rural areas where structural unemployment prevails. Afforestation projects seem to offer a solution to the problem. The employment creation aspect of forestry, especially in the rural sector, was strongly emphasised during the creation of the Forestry Commission. Thirdly, the Government also believes that the private sector should be encouraged to participate in the expansion of forestry. In this respect it allocates public money in the form of cash aids and tax exemptions to private individuals or companies who are investing in forestry.

In spite of a 70-year effort by the authorities, forestry in the British Isles has never been as important as it is on the Continent. Only a little over 5 per cent of the land is under trees in the whole of Ireland, North and South. In mainland Britain the figure is slightly higher, just over 7 per cent at the time of writing. Compare these figures with those for some other European countries: Holland 10 per cent; Denmark 12 per cent; Spain, Greece and Turkey over 15 per cent; France, Italy and Belgium about 20 per cent; Poland and Switzerland over 25 per cent; Portugal, Germany and Norway about 28 per cent; Austria and Yugoslavia 38 per cent; Sweden 56 per cent; and Finland a colossal 72 per cent. On average it takes 50 years to grow coniferous timber in mainland Britain. The figure for the whole of Ireland is between 35 and 45 years, depending on location. The gestation period for the same quality of wood in Germany and the Benelux countries is 80 years, and it is over 90 years in most parts of the Scandinavian peninsula.

Despite these islands' overwhelming absolute advantage with respect to rate of growth of trees in Europe, there is still a substantial body of economists who are not convinced about the economic viability of forestry projects. For example, a campaign was launched some years ago by a London-based research group, the Institute of Economic Affairs, to terminate state and state-aided private forestry in the United Kingdom (Miller, 1981). In its analysis the Institute pursues two basic themes. First, in the United Kingdom, forestry in general, and particularly that practised by the forestry authorities, is not a profitable venture. Secondly, whatever forestry is appropriate will be furnished entirely by decisions of landlords without state encouragement and control. Currently there are about 1.5 million hectares of private forestry in the United Kingdom for which private planters receive subsidies from the Government. Without these aids there would be little or no private forestry in this country. The Institute is in fact advocating the complete elimination of forestry in this country. The timing of this publication worried many foresters as the Institute and the ruling Conservative Government share the same ideology in that they both dislike state economic ventures and subsidies. Another reason for the hostility towards forestry in the United Kingdom is the fact that it is now a special case in the public sector investment portfolio, a case which will be explained below.

THE CHOICE OF SOCIAL DISCOUNT RATE AND PROJECT EVALUATION METHOD FOR FORESTRY PROJECTS

In economic analysis of forestry projects there are three dominant factors: the choice of discount rate; the choice of project evaluation method; and the future price of timber. An analysis for future prices will be carried out in Chapter 3. In this section attention will be focused on the first two factors. What is the official view regarding the choice of discount rate and the project evaluation method for public sector investment projects, including forestry, in the United Kingdom? In order to find answers we need to review a number of White Papers published for such purposes over the last two decades.

In a 1961 White Paper, 'The financial and economic obligations of nationalised industries' (Cmnd 1337), a number of recommendations were made on the aims and objectives of publicly owned industries. The main emphasis was on the issue that nationalised industries should 'pay their way'. In order to achieve this they must operate in such a fashion as to earn a sufficient rate of return on their capital and build adequate resources to deal with contingencies which might occur from time to time. With a low rate of return on the capital the operation of these industries will, sooner or later, damage the economy. As their activities expand this will result in either higher taxes or greater borrowing by the Exchequer in order to provide for their new developments. In this way too much of the nation's savings could be channelled into the areas where nationalised industries operate with inadequate returns. It was also mentioned in this White Paper that there was a public expectation that the products of nationalised industries should be provided cheaply. In fact the boards of these industries had been subjected to pressure from public opinion to keep their own prices down even when costs and prices were rising elsewhere. In this way they contributed substantially to efforts to stabilise prices, but on the other hand their own positions suffered. It was thought that the pressure on public enterprises should be minimised, enabling them to manage their own affairs without outside interference.

How do we ascertain that public corporations are earning adequate returns and paying their way? This was made clear in the 1967 Sub-Committee report on nationalised industries and the following White Paper, 'Nationalised industries: a review of economic and financial objectives' (Cmnd 3437). In the latter it was suggested that the Government expected nationalised industries to use the best possible appraisal technique. The discounted cash flows method which had already been widely used by nationalised industries was recommended for all important projects. It was also recommended that a test rate of discount of 8 per cent should normally be used in this method. The White Paper also mentioned that this 8 per cent test rate of discount was broadly consistent (having regard to differing circumstances in relation to tax, investment grants, etc.) with the average rate of return, in real terms, earned on a low-risk project in the private sector. In August 1969 this rate was increased to 10 per cent; in 1972 it was reviewed but not changed.

The discounted cash flows, DCF, is a simple method in which the costs and benefits generated by a project over time are discounted and then summed up to obtain an overall figure. If this figure turns out to be greater than zero, then the project in question becomes feasible, otherwise it fails, indicating that the project does not generate discounted benefits large enough to offset discounted costs. This formula is:

$$\mathrm{DCF} = \frac{B_0 - C_0}{(1+r)^0} + \frac{B_1 - C_1}{(1+r)^1} + \cdots + \frac{B_n - C_n}{(1+r)^n}$$

or

$$\text{DCF} = \sum_{t=0}^{n} \frac{1}{(1+r)^t} (B_t - C_t)$$

where

t = time, normally expressed in terms of years, 0, 1, 2, ... n
n = project's life
B_t = benefit at time t
C_t = cost at time t
r = appropriate discount rate

$\frac{1}{(1+r)^t}$ = Discounted cash flows discount factor at time t. (A discount factor matrix for various rates of interest and for various years is provided in Appendix I)

Example 1.1

Cost and benefit details of an investment project are given below. Find the discounted cash flows by using 10 and 20 per cent interest rates.

	Year 0	*Year 1*	*Year 2*
Estimated costs	250	50	30
Estimated benefits	0	200	190

10 per cent interest rate

$$\text{DCF} = -\frac{1}{(1+0.1)^0} 250 + \frac{1}{(1+0.1)^1} 150 + \frac{1}{(1+0.1)^2} 160$$

$$\begin{aligned}\text{DCF} &= -(1)(250)+(0.909)(150)+(0.826)(160)\\ &= -250+136+132\\ &= 18\end{aligned}$$

So, the project qualifies at a 10 per cent interest rate.

20 per cent interest rate

$$\text{DCF} = -\frac{1}{(1+0.2)^0}250 + \frac{1}{(1+0.2)^1}150 + \frac{1}{(1+0.2)^2}160$$

$$\begin{aligned}\text{DCF} &= -(1)(250)+(0.833)(150)+(0.694)(160)\\ &= -250+125+111\\ &= -14\end{aligned}$$

The project fails at a 20 per cent interest rate.

This method is also known as the net present value criterion. Strictly speaking there is a difference between the discounted cash flows and the net present value criteria. In cost–benefit analysis, a branch of economics in which public-sector investment projects are evaluated, the aim of the net present value is to capture all costs and benefits which are associated with these projects. Sometimes a part of the net benefit stream is non-encashable and the analyst has to estimate this and then express it in terms of money. The discounted cash flows method, on the other hand, limits itself to only the cash consequences of projects. In other words, the discounted cash flows method is a somewhat less comprehensive version of the net present value criterion. Nevertheless, some people use these terms interchangeably.

In 1967, some economists seemed to be satisfied with the choice of test rate of discount, and almost all were happy about the use of the discounted cash flows method in the appraisal of public-sector projects. For example, Alfred (1968) first calculated a typical rate of return of 6.2 per cent for the United Kingdom economy as a whole. After considering the income tax on profits he then increased his figure to 7.1 per cent which was close to the government's first proposal of 8 per cent. In view of the government's replacement of income and profit tax by a 42.5 per cent corporation tax in April 1965, Alfred finally modified his estimate to 10.1 per cent. This figure was also remarkably close to the increased test rate of discount.

However, some economists pointed out a number of problems with the choice, as well as the magnitude, of the test rate of discount. First, the test rate of discount was based entirely on the opportunity cost rate argument. Indeed, there is a substantial body of economists who have always advocated that a public project would involve the sacrifice of some other projects. Thus the proper rate of discount must be the social opportunity cost rate. This rate is defined as the one that measures the value to society of the next best alternative investment project in which funds could have been employed. Generally, those next best alternatives are sought in the private sector, and the objective behind the use of the social opportunity cost rate is to avoid displacing better investments in the private sector. If, for example, new investment projects in the private sector are earning a real rate of return of, say, 10 per cent, the public sector projects should earn at least the same rate. However, this is only half of the story. There is an equally powerful school of thought which has suggested the use of a different discount rate, the social time preference rate, which is also known as the consumption rate of interest. This school argues that the reason behind putting money in investment projects is to enhance future consumption capacity. In other words, an investment project involves a trade-off between present and future consumption. Therefore, what we need to do is to ascertain the net consumption stream of investment projects and then use the consumption rate of interest as a deflator. For a much wider argument see Steiner (1959), Feldstein (1964, 1974), Arrow (1966), Arrow and Kurz (1970) and Kay (1972). The social time preference rate is defined as a rate that reflects the community's marginal weight on consumption at different points in time. For actual derivation of this rate for a number of countries see Kula (1984a, 1985, 1986a). The economic theory suggests that in the choice of a social rate of discount the two rates, i.e. the social opportunity cost rate and the social time preference rates, should play a joint role (Fisher, 1930; Eckstein, 1957, 1961; Feldstein, 1964, 1974; Marglin, 1963a). Nowhere in the White Papers was there a mention of the social time preference rate.

Secondly, in estimating the social opportunity cost rate as a basis for the test rate of discount, the government used extremely crude measures. The Treasury specified, in a memorandum submitted to the Select Committee on Nationalised Industries in 1968, how the figure was chosen (HCP 371-III appendices and index, appendix 7). It was the minimum return which would be regarded as acceptable on a new investment by a large private firm. Obviously this commercial rate of return on private capital was adjusted neither for market imperfections nor for externalities. In other words, the stock market's view of rate of return was taken to represent the social opportunity cost rate. Moreover, a crude average rate of return, rather than the marginal one, was taken in the choice of a figure for the test rate of discount.

Thirdly, the funds that were used to finance the capital investments of nationalised industries were assumed to displace private investment rather than private consumption. In order to finance public projects, if the government borrowed the entire funds from the capital market, then the social opportunity cost rate would have been relevant provided that the private rate of return was adjusted for market imperfections and externalities. However, in most cases projects are financed by the tax revenue, a large proportion of which comes from the reduction in private consumption. The appropriate course of action in this case would be to use the consumption rate of interest rather than the social opportunity cost rate.

The strongest objection to the test rate of discount came from the pro-forestry lobby as forestry economists such as Price (1973, 1976) and Helliwell (1974, 1975) argued that the 10 per cent test rate of discount had ended the hopes of an economic rationale for forestry investment in the United Kingdom. Indeed some earlier studies on forestry in Britain had revealed that a very low discount rate was needed, of the order of 2 per cent, so that a positive discounted cash flows figure could be obtained (see Walker, 1958; Land Use Study Group, 1966; and Hampson, 1972). Some economists such as Thompson (1971) argued that in most cases it is not possible to earn more than 3 per cent compound from forestry in Britain without subsidies. However, he also maintained that in most cases this figure was higher than the return obtained in other countries of the temperate northern hemisphere.

In the discounted cash flows method with a discount rate as high as 10 per cent the power of discounting practically wipes away the distant benefits that arise from felling in forestry. In order to salvage forestry projects, forestry economists have persistently argued for the use of lower rates of interest than the test rate of discount. They defended their arguments on the grounds that unlike fabricated goods the risk of land-based investments such as forestry becoming worthless in the distant future is extremely low. Since forestry is more or less a risk-free investment, then it should be discounted at a specially low discount rate.

All these criticisms of the test rate of discount turned out to be fruitful. In 1975 an interdepartmental committee of administrators and economists was set up to review the test rate of discount and to consider its relevance to a wide range of public-sector ventures. These included the investments of public-sector trading bodies and the nationalised industries' new capital projects, which take place in an environment of changing technology and market demand. The committee recognised the fact that in public-sector investment appraisal the chosen rate of discount should reflect ideas on the social time preference rate as well as the social opportunity cost rate. The committee also stated that a justifiable figure for the social time preference rate would almost certainly be below 10 per cent.

Secondly, there was no point in imitating the private sector unless the rate of return there correctly reflected society's view of profitability which was likely to be very different from the stock market's view. Thirdly, it was hardly convincing to regard a public investment as diverting resources entirely from private investment rather than private consumption. In view of all these considerations the committee recommended a rate of 7 per cent in real terms as striking a balance between the social profitability of capital and the social time preference rate.

After the committee finished its study in late 1976, there was a great deal of discussion during the run-up to the White Paper, 'Nationalised industries' (Cmnd 7131) in 1978. The appropriate rate of discount was considered in more detail with the departments directly concerned. The main change from the original recommendation was a reduction in the figure. The original 7 per cent was the average of the 6–8 per cent range. The later data showed that the appropriate profitability in the private sector was between 5 and 7 per cent, and in the end 5 per cent was chosen. This figure was called the required rate of return.

Even a 5 per cent discount rate was too high to justify the United Kingdom's forestry programme on commercial grounds. Figures revealed by the Forestry Commission in 1977 showed an expected rate of return of about 2–2.5 per cent on 60 per cent of the Forestry Commission's acquisitions in Britain without subsidies. (Forestry Commission, 1977a). These low figures were no surprise to foresters. After the publication of the 1967 White Paper, which came as a shock to foresters, many tried to inflate rate of return figures by putting up the following arguments. First, returns from forestry fail to take into account the increasing value of land acquired for afforestation. If this factor were considered, then it would no doubt improve the profitability. Secondly, a cost–benefit analysis should not be based on current prices, which fail to take into account the historical fact that the price of timber rises. The future as opposed to current price of timber should be used in any meaningful cost–benefit study. Thirdly, recreational and environmental benefits from afforestation must also be incorporated into cost–benefit studies. Assumptions must be made about the expected visits to plantations by the general public as trees grow older and more attractive, about wildlife conservation, etc.

Others simply refused to accept the view that the United Kingdom's forestry programme should be judged on the basis of profitability, even if it was widely defined to include social and environmental factors. For example, the chairman of the Forestry Commission in 1971, Lord Tylor of Gryfe, in a speech to the Scottish Woodland Owner's Association on 4 May 1971, said:

> the Forestry Commission should not be regarded as a large commercial state enterprise but should be recognised for what it is — as the forestry authority concerned not only about the state enterprise but concerned about the propagation and development of forestry as a whole in Britain. Therefore, I consider it my mandate and responsibility to speak as effectively as I can for forestry as distinct from any particular section.
>
> *The Times*, 5 May 1971.

In 1971 an interdepartmental government team was set up to carry out a cost–benefit analysis of British forestry. In this study a wide definition of profitability was taken as the criterion. The team considered the environmental, landscape and recreational benefits of forestry along with the value of the timber produced. Also the team tried to impute shadow prices, i.e. true prices based on the scarcity value of resources, for the inputs as opposed to distorted market prices. Various sensitivity analyses were also carried out. In one, a 20 per cent premium for import saving was levied on the home-grown timber. After all these painful and time-consuming calculations to improve the profitability, the maximum rate of return from forestry was increased to only 4 per cent. The results of this study were published in 1972 (HMSO, 1972a).

Following this study, in 1972 the government revealed its forestry policy in a document in which it was stated that afforestation would continue in Britain as before (HMSO, 1972b). This document also indicated that this policy was guided by but not based on the conclusions of the cost–benefit study. With regard to the discount rate, the Forestry Commission in Britain was given a target real rate of return of 3 per cent to be used as a discount rate in the discounted cash flows method in evaluation of forestry projects. The justification for such a low rate was the unquantifiable benefit of maintaining rural life and improving the beauty of the landscape, both of which are associated with forestry. In 1973 the Forest Committee of Great Britain commissioned a study, published as the Wolfe Report (1973), to provide an argument to counter the 1972 cost–benefit analysis. In this study assumptions and omissions in the 1972 cost–benefit analysis were emphasised.

By now, the reader must have realised that the conflict which took place between various groups in the 1970s to justify or fault the United Kingdom's forestry programme was centred around the magnitude of the discount rate. Obviously, 3 per cent target rate of return makes the United Kingdom's forestry a special case, and this is not a desirable policy in the sense that it may open the floodgates for many other special cases in the future. Moreover, it is well known that a policy that permits different rates of interest, or principles, for projects will misallocate resources in the public sector. No economist who is committed to the idea that scarce resources must be allocated efficiently between sectors can be happy with this practice in which a number of rates are prevalent in the public sector. There must be one rule for all public sector projects.

2

Forestry in Some Selected Western Countries and the European Economic Community

This chapter contains basic information about forestry in Canada, the United States of America, France, West Germany, Italy, Spain, Finland, Holland and the Republic of Ireland and more generally on the European Economic Community. It is intended for economists and general readers who do not have a strong background knowledge of forestry. The chapter may be useful to non-specialists in that it explains how historic events have had an effect on the current structure of forestry in these countries. Although the history and structure of forestry are different in each country, as the reader will observe, there are a lot of problems that the countries have in common. Those with a detailed knowledge of forestry are likely to be knowledgeable about the history and structure of forestry in these countries and therefore may skip this chapter.

In many countries forests meet an important need for industrial materials, providing economic activity and employment for a large number of people (especially in less prosperous regions), and supporting activity in wood processing industries. They also play a crucial role in maintaining the ecological balance and contributing to environmental quality by preventing soil erosion and desertification. Furthermore, forests provide a base for recreational and leisure activities for urban as well as rural dwellers. In many Western industrial nations, with the notable exception of the US, Canada and Scandinavian countries, the need for forest products greatly exceeds the amount of wood obtained domestically. Major nations of Europe, such as the United Kingdom, France, Germany, Italy and Spain, are all net importers of wood. Another major wood importing nation is Japan, the largest buyer in the world. The second largest importer is the United Kingdom, and Italy comes a close third. In 1985–86 the United Kingdom imported more than £4000 million worth of wood products, the second largest item in her import bill after food. Nowadays wood has become an item nearly as important as oil for many European countries in their balance of payments account.

Because of all these reasons many countries of the world have forestry policies with a multitude of objectives. The priority for countries with a large trade deficit, such as the United Kingdom, is to expand the source of domestic timber in an environmentally acceptable way. Expansion of the forest area would also provide employment in the rural sector. An increase in the volume of wood effectively available from the forest would also allow wood processing industries to expand. In some of the already heavily wooded areas of Europe, where forestry is well organised, large numbers of jobs are directly dependent on forestry. For example, in the West German state of Baden-Württemburg it is estimated that forests provide employment for about a quarter of a million people.

Some areas of marginal agricultural land in Europe provide ideal growing conditions for trees. For example, in the west of Ireland large areas are not suitable for lucrative agriculture but are good for forestry. Ironically, most of the rural population in those areas are engaged in traditional types of farming such as sheep and cattle, and the income generated by these means is not enough to sustain families. In parts of counties Galway, Mayo and Sligo the most rural communities rely heavily on social security payments. Because of the poor farming conditions rural depopulation has always been a major problem in that part of Ireland. Since 1982, with the help of the EEC, the Irish government has been actively promoting afforestation projects in the west to prevent further depopulation. There is considerable scope in these areas for short-term cropping with a view to press-board production as well as for traditional forms of timber production.

It is a well publicised problem that, in the EEC, agricultural surpluses have been growing steadily after the introduction of the Common Agricultural Policy (CAP). Currently, the Commission is considering possible measures for the development of forestry as an alternative to agriculture. In this respect, the cost of supporting forestry production has already been seen in relation to the cost of agricultural support and that of other measures for taking land out of agricultural production.

Although there are problems which are common to most Western countries, such as over-production of food and under-production of timber, the structure of forestry and hence governments' policies are different in each country. The major problem facing the timber deficit nations of the West is to take some land out of agriculture and put it under trees. The forest-rich nations, on the other hand, want to improve the productivity of their forests without diminishing stocks. There are also environmental problems such as acid rain, which is threatening forests throughout Europe and North America.

CANADA

Canada is one of the richest countries in the world with respect to forest resources. The forests cover a vast area, extending from the Atlantic to the Pacific and from the US border to the Arctic tundra. There are about 150 species of trees native to Canada, 30 of which are coniferous. The dominant ones are various types of spruce, fir, birch, beech, maple and poplar. In the western coastal region where the climate is mild and precipitation is high, the native species, especially Sitka spruce and Douglas fir, thrive. On the eastern coastal regions red, white and black spruce, red and jack pines, birch and maple are widely distributed. The south-eastern part of the Ontario peninsula is most favourable for deciduous forests. Species that are found here cannot be found elsewhere in Canada. However, the area is heavily populated and forests are represented now only by wood lots, parks and small forest areas. The extreme northern parts of the country are either tree-less or contain unproductive woods. Between the southern and northern extremes of the country a large number of species can be found.

The forested area is about 1 400 000 square miles, or 40 per cent of the total land surface. Because of its remoteness and adverse climatic conditions, about 40 per cent of the country's total forested area is classified as unproductive. However, most of the unproductive woods provide shelter for wildlife and conserve water supplies. The productive forests cover roughly 800 000 square miles in the country. The inaccessible forests constitute a reserve for the future and their development must await the advance of satisfactory means of transportation and the expansion of the markets, especially local, for forest products.

British Columbia is one of the most thriving districts in wood and related industries. Here one harvester can cut about 130 trees in one working day. However, it has been stated that almost half of each tree felled is wasted, and replanting is not consistently done. Reed (1979) argues that the next crop of trees is going to be smaller than the last one, and it is going to take longer to grow. About 20 per cent of the area harvested each year fails to regenerate. Fire and insects take about 120000 hectares annually. In the east the adverse effects of air pollution on forest productivity are increasing.

In Canada, the wood harvesting and processing industries employ about 400000 people, and forest products presently comprise 15 per cent of all manufactured goods. The country leads the world in the export of forest products, and the annual value of shipments is about 12 billion Canadian dollars. The importance of wood in the economy is constantly growing. For example, the timber industry has doubled its capital expenditure to $15 billion in the current five-year period over the preceding five years (Hartwick and Olewiler, 1986). In order to protect forests the government also increased regeneration spending to $650 million in 1986 from $250 million in 1981. Forestry experts believe that more land under trees would be desirable, but they also indicate that proper management could halve the time needed to grow a commercial tree.

About 73 per cent of the total forest land is owned by the provincial governments. Some 20 per cent, mostly in the Northwest Territories and the Yukon, is owned by the federal government. The size of the privately owned forests is less than 10 per cent of the country's assets. The government's policy is to retain land ownership but to grant timber cutting rights to private companies. About 10 per cent of the forested public land is under timber licence. These areas, along with the privately owned forests, are in the more accessible regions of the country. Forests within the provinces are administered by the provincial governments. The federal government is responsible for the administration of forests in the Yukon, the Northwest Territories, Indian reserves and the national parks.

In the mid-1940s a Royal Commission on Forestry was appointed to study the forestry situation to make recommendations in Saskatchewan, British Columbia and Ontario.

Following this, in 1949 the federal parliament passed the Canada Forestry Act, authorising the federal government to give the provinces financial assistance in carrying out improvement work. In 1951, assistance was offered to the provinces to initiate a 5-year plan aimed at completing their work on forest inventories. After becoming confederated with Canada in 1949, Newfoundland had its first Royal Commission on forestry in 1954. The surveying and forest improvement work in this province started in the late 1950s, and the provincial government was put in charge of the development of forestry there. In Nova Scotia the forestry situation differs from that in the rest of the province as 75 per cent of the forests are privately owned. However, owners must obtain permission from the Department of Lands and Forests to fell small trees.

Canada will continue to be a leader in the wood and wood-related products industry because so much of her vast land area is better suited to forests than to anything else.

UNITED STATES

Before colonial times forests covered the entire eastern half of the United States in an almost unbroken fashion from the Atlantic right up to the central plains. When the explorers landed, they saw America as a sea of trees. The view from the mountain-tops was overwhelming as trees stretched in every direction as far as the eye could see. The early Spanish explorers coming to the northwest from Mexico were also astonished by the vastness of American forests. These early settlers cleared some land of trees in order to make room for crops, but such clearances were mere nibbles at the edges of the great forests. Trees provided ready building material and fuel, which were as important as crops to the early pioneers. As the nation expanded westwards lumber from the forests provided a major part of the building materials. West of the Great Plains, forests occurred in more isolated groups, chiefly in the mountainous areas where rainfall is heavier. Towards the middle of the nineteenth century, the area of forest cut annually was about 800 000 hectares. This figure went up to about 4 million hectares at the turn of the century. A considerable part of the cleared land eventually reverted to trees but there was a substantial time lag. The forest area of the United States has been stabilised since the First World War. Today, about 270 million hectares of forests cover the continental land mass of the United States, which amounts to 34 per cent of the land surface. It was estimated by Clawson (1979) that around the mid-nineteenth century this figure was 50 per cent. In Alaska, forested land is about 60 million hectares.

The present-day forests contain about 800 native species. This variety is due to the extremely varied conditions of climate, soil type and altitude. About 60 per cent of the nations's forests are in the eastern United States. In the north-eastern corner, spruce, fir and pine predominate. The area from New Jersey to Texas, the southern pine region, is dominated by pines. This area supplies nearly 60 per cent of the pulp and more than one-third of the timber cut in the United States. The region between New England and the Mississippi Valley abounds with various species of oak and hickory. Maple, birch and beech are widespread in the lake states and parts of Pennsylvania, New York and New England. Parts of the lake states are dominated by aspen, which is usually a short-lived species and may ultimately be displaced by maple and fir. In contrast, in the bottom lands of the south there are dense hardwood forests of oak, sycamore and maple and a variety of other trees and shrubs. In the west, high up in the Rocky Mountains up to 11000 feet (3355 metres), and in limited areas of Oregon and Washington, fir and spruce forests are common. However, commercial uses for these western forests have so far been limited, though they are of great importance for watershed protection and for recreation. Until recently these were the most heavily exploited forests in the United States. Among the more famous forests are the redwoods of northern California, which can be found on a narrow belt about 100 miles long and 20 miles wide (1 mile = 1.609 kilometres). Some stands contain more than 5000 cubic metres of trees per hectare — the heaviest stands of timber in the world.

Wood was the basic fuel throughout the US until well into the late nineteenth century, and it remained so on most farms right up to the beginning of the Second World War. One reason for its prolonged use on farms was the fact that most of them lacked electricity until the end of the 1930s. Lumber, another large consumer of wood, was dominant in the late nineteenth and early twentieth centuries. The major activity which it supported was construction, which became very depressed in the 1930s. Since the end of the Second World War, lumber output has increased again and has remained fairly high. Pulp, plywood and veneer have become major wood-using sectors in the last few decades. When all forms of manufactured wood in the United States are considered, an irregular but significant upward trend over the last 200 years is evident. Of course, different uses of wood over time require different kinds of wood. For example, lumber and plywood are made today from species whose log qualities and sizes were considered quite unsuitable only a couple of decades ago. Much of the wood used as fuel in earlier times would be considered suitable for manufacture today. Currently in the United States the annual lumber output is about 400 million cubic feet (1 cubic foot = 0.028 cubic metres). Wood for plywood, veneer and paper constitutes another 5 million cubic feet, most of which is used domestically. Under 2 million cubic feet of wood is consumed for fuel and other purposes. These figures make the US the largest wood-consuming nation in the world.

As for the ownership of forests, the situation has changed over the years. Until the turn of the last century, ownership was mainly public, federal, state or other, and the land policy was to convert the lands in the public domain to private ownership. Around 1900, concern for future timber and water supplies led to lands being reserved for national forests and to basic changes in other land policies. In the early years of this century, many federal governments purchased lands back from the private sector, which increased the size of the national forest areas. In the 1930s the federal and state governments repossessed millions of acres of private forests, mainly for non-payment of taxes. The public/private division of forest ownership has stabilised during the past few decades. Today more than 50 per cent of all US forests are privately owned. Some of the private forests are owned by the forest industry firms which have increased their holdings steadily since the turn of the century. During the last few decades there has also been a decrease in farm forest acreage.

In the United States the national forests are managed by the Forest Service under a decentralised system of operation calling for 11 regional foresters, 150 forest supervisors, and 750 district rangers. Each district ranger administers timber, water, wildlife and recreational resources in an area that frequently exceeds 100 000 hectares. The objectives of the Forest Service are to achieve the maximum potential for the national forests, to provide wood, to protect wildlife habitats, to provide and protect a water supply, and to provide facilities for outdoor recreation. The Forest Service is not anticipating any major shift in land use, at least up to the early part of the twenty-first century. The withdrawal of land from forests is also likely to be very small and mainly for transportation and urban redevelopment. On the other hand, abandonment of farmland is also likely to be very small, and some land may be shifted back into forestry. The Forest Service expects that the volume of standing timber will increase modestly over the next few decades. This trend will include some liquidation of old-growth timber, but this will be more than offset by increasing inventory from younger stands which will meet the increasing demand in sectors such as pulp and plywood without any contraction in timber stocks. The total consumption of lumber may have more or less stabilised but there may be some increase in firewood consumption.

In addition to wood supplies the contribution of forests to wildlife, recreation and watershed values is substantial. These aspects of forests are widely appreciated today, particularly as Americans have become more affluent and more urban. In the mid-1920s total recreational visits to the national forests were about 6 million; today the figure is about 200 million (Clawson, 1976). For many years recreational use of forests increased at a rate close to 10 per cent per annum. From the early 1930s to the late 1970s numbers of big game killed in the national forests increased by about 2.5 times. Over the same period the amount of forage consumed by wild animals increased by about 6 times. The amount of water flowing off the national forests has probably remained steady, but the water stored in dams or used for irrigation has increased substantially. There is doubt that public interest about the non-wood output of the national forests will continue to increase.

As was mentioned earlier, there are two distinctly different kinds of privately owned forests in the United States, those owned by wood industry firms and those owned by others. The former own about 14 per cent of the entire commercial forests in the country. Most of the firms are vertically integrated and their operation is extensive in all major forested regions; some even have subsidiaries in other countries. The other owners are mainly smallholders who have no manufacturing facilities to process their wood. The forest industry firms have achieved a much higher growth rate per hectare than other owners, mainly because of better management and the advantage they have in large-scale operation.

One problem which the private owners are facing is that of externalities. Both industrial and non-industrial owners are unable to capture the full benefits of their forests, especially with regard to water supply, protection of wildlife, recreation and aesthetic enjoyment. All forest owners suffer from trespass for these uses. Trespass for the purpose of cutting trees for fuel is on the increase. It is possible to impose an entry fee, but no charge can be made for hunting because the wild animals in forests and elsewhere belong to the state. Likewise a forest owner is unable to capture the value of the water flowing out of the forests. Policing the estate to prevent unauthorised use is a very costly matter in most cases and this makes many forest owners helpless. To make matters worse, an attitude has grown up in many localities that access to private forests should be free to the public. In the past some forest industry firms closed their property to public access; the public reaction was so hostile that today almost all firms make their lands freely available to hunters and other leisure seekers. Without a monetary reward for providing non-wood outputs, owners have no incentive to invest in the production of such outputs, nor do they have any inclination to preserve the existing facilities.

FRANCE

France has one of the largest forestry sectors, 15 million hectares, in Western Europe. Twenty-seven per cent of the land surface is under trees and almost three-quarters of this is privately owned. Like the rest of Western Europe, France was much more heavily wooded at the time of the Roman Empire, and even in the Middle Ages, than it is today. Despite her current position as a well forested country in Europe, forest destruction was a major problem in the past. Since the arrival of the Gauls, about 3000 years ago, a process of deforestation took place and by 1300 the forests covered only about 13 million hectares. Later on, the demand for wood as fuel and construction material became so great that by the beginning of the French Revolution the wooded area had shrunk to about 8 million hectares. The destruction continued until the early part of the nineteenth century but after that the area under trees began to increase slowly but surely. The two World Wars did inflict some damage on French forests but this was not as extensive as it was in the British Isles.

The most heavily forested regions of France are the south west and the mountains which ring the country on the east, the south east and the south west. The least forested regions are Brittany, Normandy, Maine and Poitou. Broad-leaved trees occupy more than 60 per cent of the forest area, and the principal species are oak, birch and hornbeam. In the Mediterranean region, small scrubby trees with poor wood quality dominate. In this area the landowners are apathetic, partly because of the constant and serious threat of fire.

The roots of modern forestry regulation in France can be traced back to the early nineteenth century. Before that the history of forestry administration was one of recurrent crises. A period of extensive felling led to shortages of wood and this in turn to tough and almost unworkable forest legislation. For example, in the seventeenth century it was feared that France would perish because of lack of timber for her Navy, which led to harsh regulations to protect and expand the forestry sector. Nevertheless, a century later the forests were again declining in quantity as well as quality. Extensive destruction in the Alpine region created many devastating torrents and floods, which helped to bring home the need for reform. France and Switzerland were the first nations in Europe to understand and evaluate the protective function of forests. In 1827 they developed a new and comprehensive forest code which included the necessity for forest research and many principles of silviculture management. Of particular importance were techniques of natural regeneration designed to restore the much abused forests.

Improvement was very slow, largely as a result of the fact that a very high proportion of forests in France were in private hands. Although some private owners learned the new techniques, private forest estates remained immune from state control and supervision and on the whole were less well managed than the public forests. Furthermore, about 70 per cent of the public forests were communal, which left only a small percentage under the direct and permanent control of the government.

In France, ownership varies considerably from region to region. In the Paris Basin, Landes and Massif Central a very large proportion of the forest is under private ownership, whereas in the Pyrenees and the Alps the state forests dominate. Most of the public forests are fairly large, but the private forests are fragmented into small ownerships. Productivity in French forests is not high but the situation is improving steadily as the owners learn new and improved management techniques.

The public forests are administered by the Service des Eaux et Forêts, which is part of the Ministry of Agriculture but enjoys a high degree of independence. The country is divided into various forestry districts, each covering up to 100 000 hectares of forest and each under a district officer and an *ingénieur.* Each forest has a technical officer and a corps of rangers. Felling decisions in each region are made by the *ingênieur.* Normally the trees are sold standing, at auction, and are then felled and hauled by private contractors who also carry out road building and drainage. In evaluation of forestry projects the Service des Eaux et Forêts mostly uses a 3 per cent discount rate, a figure which is somehow linked to the biological growth rate of a typical forest.

There is no grant aid scheme for private foresters. Tax relief is the only major incentive. Agricultural land in France is taxed, but if it is planted with trees the owner is exempt from these taxes for 30 years. Another form of incentive is low-interest credit for up to 30 years at 0.25–2.5 per cent, depending on the size of woodland. In certain regions, such as Brittany, young trees are provided free of charge. The Service des Eaux et Forêts provides advisors for each region and the first advisory visit is free of charge; a charge per day is levied for every succeeding visit. In some parts of France, such as Brittany, woodland owners are keen to form co-operatives which are encouraged by the Service des Eaux et Forêts. Central and local government grants are also available to these co-operatives for administration and the collective purchase of machinery.

WEST GERMANY

Germany is a more densely forested country than France, the forests occupying 7.3 million hectares, which is 29 per cent of the land area. After the redrawing of the European map at the end of the Second World War, Federal Germany lost 2.75 million hectares of forests to East Germany and about 3 million hectares to Poland and Czechoslovakia. Forest productivity is very high in Germany. Although she has about 50 per cent less forest per capita compared with France, the timber yield is about the same on a per capita basis.

During the time of the Roman Empire about 75 per cent of German land was under trees. As the population grew it took its toll on the forests. However, forest decline was halted towards the end of the nineteenth century. After the Second World War about 7 per cent of the German forests were felled to supply fuel to destitute cities and reparations to the French and British.

The proportion of land in forests generally increases from north to south. The northern coast is the least forested region of Germany. The best wooded areas are the Alpine slopes (below the timber line), Baden-Württemberg and Brandenburg. About 70 per cent of German forests are coniferous, and spruce and Scots pine are the dominant species. Silver fir and larch are also important. Broad-leaved forests are dominated by oak and beech. In the past the mixture was in favour of hardwood. Two factors have caused the transition from deciduous to coniferous. First, the most fertile land was occupied by broad-leaved trees which had to be cleared in order to make room for the agricultural expansion to support the growing population. Secondly, in the forest development of the nineteenth century, conifers were given preference because they grew faster and straighter than hardwoods. Spruce was particularly prized and it was planted all over Germany.

Fearing timber famines, many German communities between the thirteenth and sixteenth centuries laid down strict rules prohibiting overcutting. The systematic development of a forest tradition started after the Thirty Years War. Rules were established to govern felling, haulage, grazing and activities of wood-using industries in order to avoid waste. In order to ensure permanent local supplies many communities adopted various types of sustained-yield management. The scientific bases of silviculture were developed during the early nineteenth century by G. Hartig and H. Cotta. After surveying many large forests they made recommendations to organise them on a sustained-yield basis. This involved new techniques of cutting, regeneration and thinning.

After the First World War a back-to-nature movement became prominent which also advocated that German forests should be treated as a national heritage rather than as a great wood factory. Emphasis was placed on aesthetic and recreational values. This movement became even more prominent during the Nazi era. Its members held the view that the ideal system was the removal of individual trees in a manner that would affect the forests as a whole as little as possible. This somehow tied in with the dominant political philosophy of the time.

Today ownership is almost equally divided between the public and private sectors. The German forestry authorities have extensive powers to enforce their forestry policies. These are implemented by the Bundesministerium für Ernährung, Landwirtschaft und Forsten und Länder. It is illegal to clear forests, and any felling work done must be accompanied by regeneration. The government's forestry policies are for the expansion of the productive forests, the improvement of forest management techniques, conservation, and promotion of fundamental and applied research.

There are no direct grants given to the private sector for planting or management. Assistance is available by way of establishing and administering forest-owner co-operatives. The forestry authorities also provide advisory services, education and training. The co-operatives provide plans for members, market wood and purchase co-operative machinery at a subsidised price. These subsidies and administrative costs are borne by the government. There is no compulsion for owners to belong to co-operatives but they are more than happy to become members.

ITALY

The area under trees is about 5.7 million hectares, about 20 per cent of the land surface. However, only 3.9 million hectares of it can be described as productive forests, which constitutes only 13 per cent of the total area. There are basically three forest zones in the country:

(1) Mediterranean zone from sea level to about 800 metres, which is dominated by *maquis*. The productivity of forests in this zone is low; annual growth is only about 3 cubic metres per hectare. However, *maquis* is of great economic importance to communities as a source of charcoal. In this region there are occasional good pine forests, especially in the north.

(2) Regions of the Apennines and Alps up to 1400 metres altitude are dominated by deciduous hardwoods, especially oak, chestnut and beech. Chestnut (*Castanca dentata*) forests were almost entirely planted for fruit, mostly in the nineteenth century, and they are getting rather old and vulnerable to disease. Coppicing is widely practised on this species for a variety of products such as sticks, baskets, packing and binding materials, poles, clapboards and house beams.

(3) Alps above 1400 metres which are dominated by softwood species up to the tree line, sometimes mixed with beech.

During the time of the Roman Empire fertile land was used for agriculture to support the population, The Barbarian invasion caused the abandonment of these areas, and a succession of forests took possession of the land in the third to the fifth centuries. In the following centuries resettlements and increasing population put great pressure on land, which necessitated the replacement of woods by fields. However, some forests on monastic lands survived the destruction, and monks and others working to preserve their forests actually became the founders of modern silviculture in Italy. When urbanisation began to take shape in many areas, the importance of forests as protection against landslides and flooding was recognised. Forest products such as wood, chestnuts, acorns, fodder, cork, mushrooms, etc. were very important in the economies of rural as well as urban communities. In the fifteenth century when small principalities were established, rulers sought to form forest domains and passed laws. The Venetians were particularly energetic in passing laws to protect oak and spruce forests, which were important sources of timber for their navy. However, forest laws established by the rulers were, most of the time, unenforceable. When the country was united in 1870 it faced a serious forest crisis.

Forestry policy in Italy aims to achieve a number of objectives such as flood control, soil conservation, improvement of productivity of existing forests, improvement of pastures by creating shelter belts, afforestation with a view to increasing domestic timber resources, creation of employment in rural areas, maintaining historic forests and carrying out fundamental and applied research in forestry. In recent years industrial planting in the north, especially in the Po plain, has been quite extensive. Poplar, which yields high annual output (up to 30 cubic metres per hectare), is the most favoured species. In the south, eucalyptus is widely planted.

SPAIN

Although in Spain the total area under trees is about 30 per cent, the exploitable forests cover 6.5 million hectares which is only about 13 per cent of the land surface. As in Italy, the forest zones of Spain can be divided into three groups.

(1) The northern region and the Pyrenees, dominated by evergreen species. Stands of pure fir can be found in valleys in Galicia, Astruies, Navarra and Catalonia. At lower altitudes beech, chestnut and oak are important.
(2) In the central Iberian region the mountain vegetation is mostly dominated by evergreen species, sometimes mixed with beech and oak.
(3) The Mediterranean region, due to dryness, is mostly treeless. In parts the area is almost like a desert where the climate will not allow tree species to flourish.

In general, conditions in the Spanish forests are not good. At the beginning of the Christian era, it was said that a squirrel could have gone from south to north on the tree tops. This is hardly the case today.

Wars between Muslims and Christians had a devastating effect on Spanish forests. Various conservation laws decreed on several occasions in the name of the Catholic kings failed to bring satisfactory results. Today the Spanish Forest Administration is striving to introduce management methods designed to ensure the expansion of forests.

About a third of the forests belong to the government and two-thirds to the private sector. Domestically produced timber is far short of meeting the demand and Spain is a net importer of wood. The country is about 65 per cent self-sufficient in sawnwood and about 90 per cent in pulpwood. The gap between demand and supply of the domestically available wood has an important bearing on the balance of payments. However as far as self-sufficiency is concerned, Spain is less worse off than many other European countries.

The objectives of Spanish forestry policy are three-fold.

(1) An orderly and sustained utilisation of the country's wood resources.
(2) Protection, restoration and strengthening of the country's natural environment by improving and expanding forest resources.
(3) The promotion of recreational and cultural activities with the help of the forestry sector.

The institutional framework regarding forests and conservation is comprised of a body of public institutions and legislation. The most important laws in force are the law on the promotion of forest production; the law on state- and community-owned forests; laws to prevent forest fires; and the law on mountain agriculture. The main institutions are the National Institute for Nature Conservation, the General Directorate of Agrarian Production, the General Directorate of Agrarian Industries, and the National Institute for Agrarian Research, all of which operate under the Ministry of Agriculture, Fisheries and Food. Apart from these there are the schools of forest engineering, one higher and three intermediate, under the Ministry of Science and Education.

The country has a National Reforestation Plan with long-term objectives such as decreasing the current deficit of domestic wood; contributing to the recovery of desertified, unproductive areas; reclaiming areas lost due to fires; improving the hydrological system and preventing soil erosion. During 1983–84, 137000 hectares were reforested which had an investment value of about 7000 million pesetas, and 44000 hectares of land were acquired at a cost of 1300 million pesetas. One problem which is worth mentioning is the identification of forest areas and a clear definition of ownership rights. This task has always been given priority in forest administration and legislation. Currently the total number of estates lacking demarcation exceeds 6000, more than 3.5 million hectares, and some 8000 forests (about 4.5 million hectares) are still lacking boundary markings.

FINLAND

Finland is the richest country in Europe with respect to forest resources. The productive forest area is 22000 million hectares, about 72 per cent of the land surface. The volume of standing timber amounts to over 1500 million cubic metres. Although the domestic consumption of wood per capita is very high, Finland is one of the largest timber exporting nations of the world. The great size of the forests in relation to the population allows large surpluses to be exported. The ease of transport also makes felling and haulage inexpensive compared with costs in other countries. The close proximity of the sea ports on the Baltic coast is another important factor.

The size of the privately owned forests is about 14000 million hectares, which makes the division of ownership between private and public sectors 64 and 36 per cent, respectively. In the past, forest owners had been free to manage their forests much as they pleased, but sawmills had been strictly regulated. Steam-driven sawmills were prohibited until the middle of the nineteenth century. By then the government had realised that such restrictions were not in the best interests of the economy because they effectively promoted inefficient mills. With the lifting of the restrictions widespread cutting took place, mainly for the export markets. Without restriction on cutting, the fear of devastation became acute. A law forbidding ruthless cutting was passed in 1886 but it was not effectively enforced. In 1917 a new law was passed prohibiting all cutting that would endanger natural regeneration. To enforce this law a commission was established in each county. The enforcement of the law was perhaps the single most important factor which protected forests in Finland. In 1928 the cutting prohibition was extended to young, vigorous, broad-leaved forests.

Finland lies at the limit of human habitation. About one-third of its total length lies north of the Arctic Circle. It is a flat country, most of which is below an altitude of 400 metres. Despite its sub-arctic location the climate is relatively warm because of the North Atlantic Drift bringing in warm water to the Atlantic coast of the Scandinavian peninsula, leading to the frequent penetration of warm winds from the south west. Coniferous species dominate the scene in this country, but in the south west, temperate mixed forests are common. The length of the growing season is between 120 and 170 days and the time needed to grow a mature stand varies between 80 and 200 years, depending on location and species.

The basis for the planning of forestry has been the systematic forest inventories which started as early as the 1920s and are still continuing today, covering the whole country. At present there is more wood in the forests than ever before. Over the last few decades harvesting methods have improved substantially and as a result of this the total number of people employed in forestry has gone down. Permanent, professional forest workers number about 30000; about 15000 seasonal workers are engaged in harvesting.

The Ministry of Agriculture and Forestry is mainly responsible for the development of forestry in Finland. It is the highest forestry authority, in charge of preparing legislation as well as planning and supervision of forestry as a whole. The duties of the National Board of Forestry, which operates under the Ministry, are the management, protection and utilisation of the public forests, supervision and promotion of private forestry, management of national parks, and execution of the law concerning forest improvement. Finnish forest owners also have various organisations, one of which is the Central Association of Finnish Forest Industries founded in 1918. Its aim is to improve commercial and operating conditions of forest industries. Another organisation is the Central Union of Agricultural Producers, which negotiates with the buyers the price and the quality of timber output. Workers who are employed in the forestry sector belong to the Finnish Rural Workers' Union. Working conditions and wages are determined by the Union and the Employers' Association of Finnish Forest Industries.

Forestry has been of exceptional importance for the development of Finnish society and economy. On a general level, the national objectives of Finnish forest policy are set out in the Law Concerning Forest Administration 1966, and in the drafting of a timber production programme in the early 1960s. Early in 1985 a programme known as 'Forest 2000' was presented by the Economic Council which is subordinate to the Council of State. The objectives of the Forest 2000 programme are support for the general aims of economic policy; balanced long-term development of the various forms of forest use; full utilisation of forest lands; improvement in profitability in all sectors of forestry; safeguards for operational continuity in forestry and for the supply of timber to the forest industries; and the creation of favourable conditions for a viable forest industry. Since the mid-1960s an increasing part of usable wood has been left in the forests as an uncut timber reserve. The Forest 2000 programme stresses the need to move away from this practice.

Today there are about 350000 private forest holdings in Finland. Over the last few decades there has been a structural change among private forest owners as more and more forests have been transferred from farmers to town residents. The change in private ownership and the growing common interest of non-professionals has been reflected in the national forestry policy. Minimum standards governing the long-term utilisation of the country's forests are contained in the Law Concerning Private Forests 1966, a document which prohibits the devastation of forests and lays down obligations concerning the regeneration of forests. Investments aimed at improving forest yields are voluntary. The organisation of private forestry comprises two central boards, 19 district forestry boards and a large number of forest management operations. Forest owners have also been active in the commercial and production sectors as they have several production plants.

In Finland forestry and forest industries account for about 10 per cent of the national product, and the share of forest industry in the net income from exports exceeds 50 per cent. It was estimated by the National Board of Forestry that the secondary effect of forestry on the national economy was about twice as great as its direct influence. Forests are also increasingly being seen as an inflation-proof investment by the general public. All in all, in forest management, the Finns have a very high reputation throughout the world. Their abilities range from the fundamentals of forestry to the finished products of the forest industries. The basis for this is easily found in the long history of practising forestry and forest industries and of practical experience; in the high level of forestry research; in the systematic development of forestry education, and in the flexible structure of forestry institutions. Because of all these factors, Finnish forestry researchers and economists are in great demand for international projects around the world.

HOLLAND

Holland is the most densely populated country in Europe and also one of the least forested with only about 340 000 hectares of forest area, which amounts to 10 per cent of the total land surface. The country was heavily wooded at the time of the Roman Empire, but by the sixteenth century it was almost completely deforested. From then on a very slow process of afforestation has taken place, mainly for recreation and hunting purposes. The early plantations were mainly deciduous or mixed, but in the nineteenth century the trend shifted towards conifers and today they constitute about three-quarters of the forest cover. The proportion of conifers is highest in the heavily wooded eastern provinces. They are also favoured for coastal dune reclamation and erosion control. Outside forests, willows are planted extensively along most of the rivers, and poplars are planted alongside the roads.

During the Second World War the cutting was excessive as the bombing of homes and industries created an enormous demand for wood. After the war, as the Dutch economy gained ground, an increasing proportion of the rising demand was met by imports. Today the country is only about 7 per cent self-sufficient in wood. The government aims to increase this rate to 25 per cent by the year 2030.

About 35 per cent of the forest area is privately owned. Since the mid-1970s the government has been trying to stimulate private afforestation projects by giving grants. These are available for plantations greater than 5 hectares. The subsidy amounts to 80 per cent of the investment costs and in addition up to 95 guilders (1983) per hectare is paid for management operations. Forestry co-operatives are also encouraged by the government, as subsidies covering management costs of up to 80, 60, 40 and 20 per cent are available in the first, second, third and fourth years, respectively. In order to be eligible a co-operative must contain at least three owners with at least 50 hectares each. However, despite these subsidies the financial results of many private foresters have been negative and some of them have been forced to sell their estates, mainly to state or conservation groups (Filius, 1986).

If private owners open their estates to the general public, they receive an additional cash grant from the government. The reason for this is that if the private owners closed their forests to the public the pressure for recreation on state forests would be enormous. This grant is currently 95 guilders per hectare per annum, which covers the cost of maintenance and accessibility. Forests that are near the population centres receive higher grants. Also, when the owners provide footpaths and parking spaces, the costs are fully subsidised. Various other grants for nature conservation projects, landscape gardening and building forest roads are also available from the government. Apart from these there are generous tax exemptions. Profits from forestry operations are exempt from income and corporation taxes. With regard to property tax and death duty, the taxable base for most forests has been reduced by as much as 75 per cent in recent years.

REPUBLIC OF IRELAND

Unfortunately, despite the very favourable climatic and soil conditions for forestry, the Republic of Ireland is the poorest member of the EEC with respect to timber resources: only about 5 per cent of land is under trees. The forestry history for the period under British administration until the founding of the Irish State in 1922 was explained in Chapter 1. Since its establishment the Irish State has played a dominant role in organising and facilitating the afforestation programme. Until the 1980s progress was very modest, but since then greater efforts have been made to step up the afforestation. Currently the state body which is in charge of forestry is the Forest and Wildlife Service which operates under the Department of Fisheries and Forestry.

The aims of the Forest and Wildlife Service are as follows:

(1) to promote sound national forest and wildlife policies;
(2) to promote a pattern of land use so adjusted that forestry and wildlife will yield the greatest benefits, taking into account national, economic and social policies, including environmental, industrial and employment considerations;
(3) to develop a comprehensive programme for wildlife conservation.

Specifically with regard to forestry, the main activities of the Forest and Wildlife Service can be grouped into two distinct programmes as forest development and state forest management. The former relates to the extension of the forest area by state and private planting, the latter to state forest management, i.e. the management of the existing state forests.

The Forest Development Programme aims to increase the overall forest area at a rate which will be consistent so as to ensure that the rate of increase will produce a balanced yield. It should enable national wood production to be expanded systematically so that home and export markets are captured at competitive prices with an acceptable return on investment. All these should have due regard to the conservation, amenity and recreational values. The forest development programme also aims to foster private forestry in harmony with state forestry on suitable lands and to encourage good forest management. The objectives of the state forestry programme are deemed to be met by planting 10000 hectares per annum on land on which the average yield class is not less than 14 cubic metres per hectare.

The objectives of the State Forest Management Programme are to produce wood in state forests in the most economic manner and to manage land in the possession of the Forest and Wildlife Service to best overall advantage with full consideration for amenity, recreational and conservation values. In other words, the management objective is to maximise the expected net income from wood production in the state forests.

In November 1984 the Irish Government established a Forestry Review Group with a view to scrutinising the whole range of forestry activities in Ireland. This stemmed from a government statement in the National Plan that the nation's forests should be developed to the maximum advantage. In June 1985 the group submitted a report to the Minister for Fisheries and Forestry and made a number of recommendations, the most important of which was the transfer of 1 million hectares of land from low-intensity agriculture to forestry. It was noted that the family farm income on this land was very low, and return on capital, after allowing for labour cost, was negative. If planted with trees, such land would be capable of yielding, on a net discounted revenue assessment, a return well in excess of 3 per cent over inflation. The group also recommended that the Forest and Wildlife Service should be restructured, enabling it to operate more freely and to use commercial criteria in its operations.

About 95 per cent of woodlands in Ireland are coniferous, and ownership is divided between public (80 per cent) and private (20 per cent). The size of the private sector's annual planting was negligible in the early stages of forest development around 1922. The first target figure of 4000 hectares per annum was established by the government in 1933. In 1948, a more ambitious target, 10000 hectares per annum, was adopted. Initially, a considerable number of North American and European tree species were used experimentally in plantations, but the government eventually concentrated on two north-west American species: Sitka spruce, which was found to grow successfully on most mineral soils and on some peats, and lodgepole pine which was capable of growing on poor soils. There have been continuing problems with lodgepole pine related to genetic variations, insect pests, disease and stem formation. These problems, together with improvements in silviculture techniques, have recently led to a further shift in favour of planting Sitka spruce in most locations.

In 1980 a notable development took place which is likely to affect the Irish afforestation programme to a considerable extent. As part of the EEC's regional development project a forestry development scheme, which is widely known as the Western Package, was introduced for the purpose of stimulating tree planting in the 12 western counties of Ireland. This scheme is aimed at the afforestation of land which is marginal for agriculture but suitable for forestry. Under this scheme full-time farmers can qualify for a grant of 85 per cent of the approved costs, subject to a maximum of £800 (IR) per hectare. The recoupment to others is 70 per cent of the approved cost which is also subject to a maximum of £800 (IR) per hectare. Free advice is also available from the Forest and Wildlife Service. The Western Package established a target figure of 25000 hectares over a 10-year period initially. The uptake was disappointing to start with but since 1985 the situation has changed considerably when applications by the pension funds, insurance companies and investment banks started coming in.

The planting grant which is available outside the Western Package is £309 (IR) per hectare maximum. Apart from planting grants the owners receives tax concessions which are aimed at non-taxation of income from forestry. As in the UK the profits from timber sales are basically exempt from income and corporation taxes. However, unlike in the UK, Irish investors in forestry cannot offset their costs against other earnings. Growing trees are not subject to Capital Gains Tax but the land on which they grow is subject to tax. Death duty or inheritance tax has now been replaced by a new system of Capital Acquisition Tax which is applicable to all transfers of property by gift or inheritance. Land in agriculture and forestry receives special treatment under this tax, being valued at only 50 per cent of its open market value.

THE EUROPEAN ECONOMIC COMMUNITY

The needs of the 12 member states for wood greatly exceed the currently available output from the Community forests and as a result the EEC is the world's largest importer of forest products. Net imports into the Community amounted to about 17 000 million ECU (European Currency Unit) in 1984 and the situation will not improve much in the near future. This huge deficit is largely due to the insufficiency of the wooded area. Only about 20 per cent of the total land surface in the EEC is under trees. Another factor which contributes to the problem is the under-utilisation of existing forests, some of which are totally unproductive. Despite its overall deficit the EEC exports about 2 million tonnes of paper and board per year and is a net exporter of furniture. The Community believes that an increase in the supply of wood is likely to sustain more activity in the wood processing industries within the EEC.

Whereas it has the Common Agricultural Policy, the EEC does not currently have a formal forestry policy. However, the EEC is in the process of preparing a Community Action Programme. There are four main reasons that have compelled the EEC to devise such a programme. First the persistence of the agricultural surpluses can be only a temporary phenomenon. Secondly the inevitable reduction in agricultural output will lead to a search for alternative crops including forestry products. The Community's huge trade deficit in wood gives an added incentive to increase timber output. Thirdly, urgent action is needed to stop worsening destruction of European forests by atmospheric pollution and fire. The former is particularly acute in the north and the latter in the south. Fourthly, there is a need to maintain and expand economic activities and employment in the rural EEC. In the past the EEC made a number of proposals for the development of forestry. In 1979 the Commission proposed a resolution on the Community's forestry policy and the setting up of a Standing Forestry Committee, and in 1983 it proposed objectives and lines of action for Community policy regarding forestry and wood-based industries. So far no decision has been taken by the Council on any of these proposals.

Despite the absence of a clear forestry policy, in the past the EEC invested heavily in forestry in the context of its other policies. Between 1980 and 1984 about 470 million ECU were committed to forestry projects in Europe and in developing countries. For example, the Irish Western Package, mentioned previously, is one of many such investments. The EEC has also been promoting forestry in the countries of Africa and the Carribean and in the Pacific region, who are supplying the Community with tropical wood. The Community has co-operation agreements through the Lomé Convention with some of the exporting countries. Over-exploitation of tropical forests has grave consequences for climate, agricultural production and supply of wood. The EEC has a vested interest in all these matters.

The extension of the forest area in the Community in an environmentally acceptable manner appears to be a priority issue. This is particularly important at a time when the Community has huge agricultural surpluses and a very large deficit in timber. The EEC believes that an incentive package including planting grants as well as tax reliefs may be a good way to start the ball rolling. It has been observed that similar packages that already exist in some member states give satisfactory results: Scotland is a good example. However, this can only be a short- and medium-term strategy. In the long run, the objective of agro-forestry action must be to develop an activity which is self-sustaining and does not require substantial subsidies. Another problem that must be solved is the absence of a steady income in forestry. Indeed, there is a considerable time gap between planting trees and receiving income from thinnings and felling. This is quite a deterrent factor especially for low-income farmers. In order to lure them into forestry they must be provided with a regular income similar to that from farming.

The improvement of productivity in existing forests is another key area. A considerable proportion of the Community's forests is unproductive. The EEC believes that a large increase in output can be achieved with relatively small additional investment. About 60 per cent of the Community's forests held in the private sector are characterised by problems of small size: the average holding is about 8 hectares. Forests under public-sector management consist of larger plots, and the infrastructure facilities are better than those of the private forests. Extraction of timber from the remote forests is costly, and private owners are unwilling to harvest timber in such places. Another factor is the lack of demand for wood in remote areas, so extracted material requires costly transportation to places where demand exists. This situation could be improved by the creation of forestry associations for the owners. As explained earlier the co-operative movement in Finland has given very satisfactory results. The EEC is ready to help with the infrastructure problems and the creation of woodland associations where they are needed. In aiming to improve productivity it is also necessary to support the expansion of wood processing industries. Measures to improve standardisation of forest products would be a useful step to encourage the development of the timber trade.

The EEC believes that expansion of the Community's forests and the improvement of productivity in the existing ones must be carried out in an ecologically acceptable manner. There is some evidence of strong feeling in some regions at the spread of conifers at the expense of broad-leaved forests. If a grant scheme is to be implemented to encourage afforestation, it should make provision for greater planting of deciduous trees. In any forestry policy the production of timber should not be the only objective. It must be accompanied by the safeguarding of the long-term fertility of the soil, the regulation of the water cycle, the conservation of the fauna and flora, the preservation of the landscape and the provision of recreational facilities for the general public. Unrestricted access to forests may help to bring about greater public understanding of woodlands and nature in general. However, it also increases the risk of damage. The Community believes that the aim of a forestry action programme must be to promote the opening up of forests to the public with good management to reduce the risk of damage to trees.

Another area that requires urgent action is the damage caused to woodlands by atmospheric pollution. The destruction has reached considerable proportions, especially in Germany and France. About 50 per cent of the German forests and 40 per cent of the woodlands in eastern France are damaged. The future economic development of these regions is dependent on forestry and tourism. Although the causes of the damage and the mechanics involved are not completely understood, it is generally agreed that atmospheric pollution is a major factor. The pollution is caused by emissions from car exhausts and coal-fired power stations. The EEC has already made proposals to reduce pollutant emission from cars and it envisages making further proposals on speed limits and on emissions from diesel-powered cars and from trucks. It is believed that implementation of these proposals would considerably reduce atmospheric pollution and the damage to forests. A Commission proposal to establish a regular inventory of forest damage and a network of observation posts for measurement is making only slow progress in the Council.

The Commission has also proposed measures aimed at reducing other risks such as fire and insect damage to the Community forests. In relation to the former, the EEC is willing to improve the fire-fighting capacity of member states. However, it is doubtful that this issue will become a priority case in the Council. Biotic damage caused by insects or disease poses another challenge to forests, and the Standing Committee is consulted whenever a potential danger is perceived. One important measure would be the adoption of a code of conduct on genetic impoverishment, a factor which is partly to blame for the weakening resistance of European forests.

Development of a Community action programme for forestry will require additional efforts in forestry research, statistics and information. A considerable research effort has already been made in the Wood as a Renewable Raw Material Programme, which has been extended for another 5 years. Other programmes exist in the areas of agriculture, energy and environment which include forestry. However, the EEC believes that the existing research effort is inadequate in view of the ground that needs to be made up and the scale of the requirements. Likewise, statistics and other information on forests are also inadequate for the needs of a Community action programme. All in all, efforts are needed to provide the Community with research findings, data and information in all relevant areas.

3

A Cost-Benefit Analysis of Public Sector Forestry in the United Kingdom Using Ordinary Discounting

In this section a cost–benefit analysis will be carried out on the basis of a typical 1-hectare plantation of Sitka spruce in Britain. The discounted cash flows method will be used with 10, 5 and 3 per cent discount rates.

For the purposes of economic analysis the growth of trees is normally quantified in terms of volume, although it is possible to measure it by means of height, weight or dry matter. There are two important criteria in the measurement of volume, current annual increment and mean annual increment. The former represents the annual rate of increase in volume at any point in time and the latter the average rate of increase from planting to any point in time. Figure 3.1 illustrates the behaviour of these two rates. The current annual increment curve cuts the mean annual increment curve at the maximum point. For *Ceteris paribus*, if, in an even-aged plantation, felling and replanting were repeatedly carried out at this point, then this maximum average rate of volume production should be maintained for ever. Needless to say, differences in rates of growth occur within the same species on different sites. The growth figures also vary from species to species. In British conditions the annual growth can be as low as 4 cubic metres per hectare for some hardwoods and up to 30 cubic metres per hectare in the case of some softwoods.

Yield classes are created by splitting the growth range into steps of 2 cubic metres per hectare per annum, and normally even numbers only are used. For example, a stand of *yield class 16* has a maximum mean annual increment of about 16 cubic metres per hectare (m^3/ha) i.e. greater than 15 but less than 17.

There are a large number of species which can grow in the

Figure 3.1: Patterns of volume increment. MAI = mean annual increment; CAI = current annual increment

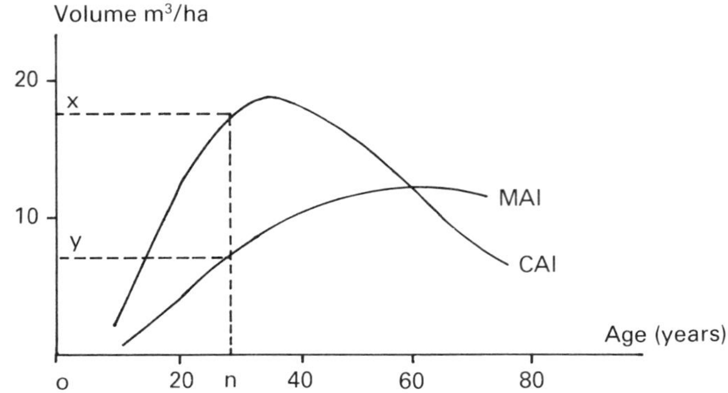

Figure 3.2: General yield class curves for oak

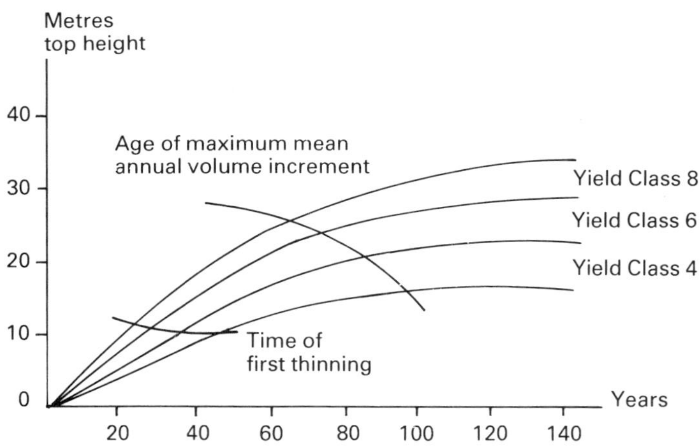

Source: Forestry Commission (1971).

British climate. Softwood species that are well established in the British Isles include cedar, cypress, fir, hemlock, larch and spruce; hardwoods include ash, beech, birch, chestnut, elm, oak, poplar and sycamore. General yield class curves for one hardwood, oak, and one softwood species, Sitka spruce, are shown in Figures 3.2 and 3.3, respectively. The range of

Figure 3.3: General yield class curves for Sitka spruce

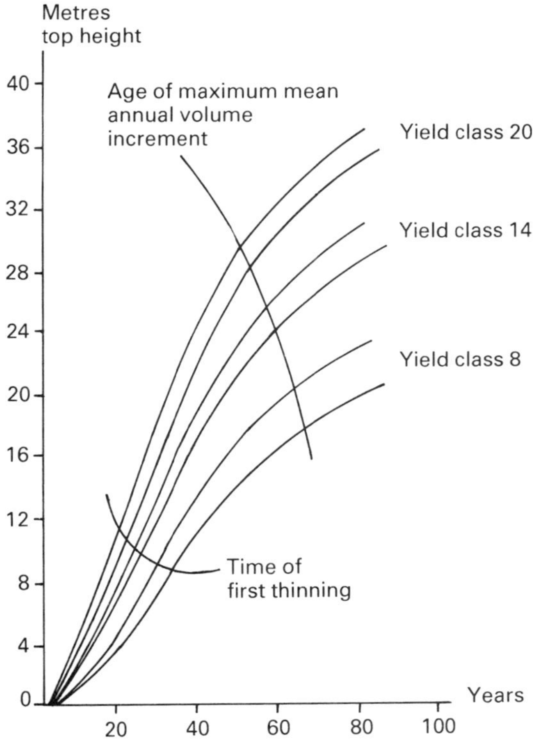

Source: Forestry Commission (1971).

maximum mean annual increment is between 4 and 8 m^3/ha for oak and between 6 and 24 m^3/ha for Sitka spruce. Because of its high yield and relatively short gestation period, Sitka spruce is commonly planted in the British Isles by the official forestry authorities as well as by private individuals.

One of the advantages of planting, as compared with natural regeneration, is that it allows foresters to select the species that will achieve the aimed objectives. The choice of species is a crucial factor in the success of any afforestation programme. In addition to economic considerations the forestry authorities in the British Isles plant trees for reasons of water conservation, prevention of landslides, climate improvement and providing protection from winds. In each of these cases different species may be required. Foresters also take into consideration the influence of particular species on soil fertility. Species which are likely to impoverish the soil in a particular area are carefully avoided. The protection of land is as important a factor as the production of timber. Of course climate and soil types in a locality are the natural constraints on the choice of species. Where conditions for tree growth are favourable, there may be a wide range of choice and many different species can be considered. Where conditions are unfavourable, the choice may be limited to a few species only. The natural distribution of trees also follows this rule. At high altitudes and high latitudes most natural forests are composed of a single species or a mixture of very few species.

The climate of the west coast of Britain is closer to that of the west coast of North America than to that of continental Europe. Districts with high rainfall and a humid atmosphere have proved especially suitable for growing the American west coast conifers such as Sitka spruce and noble and grand fir, and therefore these species have been planted more frequently in the west. European species such as Norway spruce and some hardwood trees are more suited to the east coast of Britain. Chestnut and Corsican pine, both of which come from the Mediterranean region, are valuable species in southern Britain but are seldom seen in Scotland or northern England. Scots pine, on the other hand, yields greater volume in Scotland than anywhere else in the British Isles.

Sitka spruce (*Picea sitchensis*) now dominates the British scene and is likely to become even more widespread in the future because of the good results obtained in its plantation and management so far. It is native to British Columbia and to southern Alaska. It thrives even in sub-arctic conditions, growing right up to the timber line at the edges of glaciers. It grows very quickly once it reaches about 3 metres (10 feet). If allowed to, it can grow for centuries and in favourable conditions can attain a height of over 60 metres (200 feet).

Sitka spruce was introduced to Britain in 1831, but attracted little attention until after the end of the First World War. It is usually turf-planted on ground previously ploughed and often artificially drained. It tolerates poor soil but is intolerant of competition from heather. The most attractive aspect of this species is its high and quick yield of timber. The timber quality of Sitka spruce grown in the British Isles is not, however, as good as that of timber grown in North America, but it makes a very good pulp. It is also used widely in the coal mining industry as pit props, and in house building mainly as interior construction material.

It is important to bear in mind in afforestation that relying entirely on a single species such as Sitka spruce is not a prudent decision. After all trees are living things and like any other living thing they are vulnerable to diseases, especially on non-native soil. An epidemic like the Dutch elm disease which occurred some years ago may inflict heavy damage. Just as a financial investor tries to diversify his portfolio, a prudent entrepreneur should plant more than one species. Indeed, Sitka spruce can be successfully planted with other types such as Norway spruce, lodgepole pine and Scots pine. Spruce is currently suffering from infestation by the Great Spruce Bark Beetle, which is inflicting damage on plantations in Wales and western England. Apart from this and a few other problems of a minor nature in the past, nothing very serious has so far occurred.

Our cost–benefit analysis is carried out for Sitka spruce, yield class 20, on the basis of a 50-year single rotation which typifies the gestation period in many locations in the United Kingdom. The calculations are carried out for 1 hectare, and 1978/79 prices are used throughout.

COST OF PLANTATION

There are a number of cost items. These are as follows.

Ground preparation

Planting Sitka spruce on a good location is plain sailing and it can be carried out without much soil preparation. The young plants take root quickly and need little attention. Lush and fertile lowlands are ideal locations for many cultivars. Areas of heavy and wet clays require more work by way of ground preparation. The usual technique is to raise grass mounds with spades or by ploughing and to plant through them so that the roots spread between the grass on the surface and the grass on the inverted sod. In places where the land is covered with scrub and heather, a mechanical chopping is necessary before ploughing.

For methods of controlling heather see Forestry Commission (1977b).

Afforestation on blanket bog requires a different method of ground preparation. An intensive drainage programme is required before planting. In this, by following natural drainage channels, drains are opened. On excessively wet ground, cross-drains are made at 30 foot (9.1 metres) intervals and they are deepened to 36 inches (0.9 metre)

The costs of ground preparation in various locations are shown in Table 3.1.

Planting

In most peaty locations the trees are hand-planted along the continuous peat ribbons which are turned out to one or both sides of the drain. After planting, a small amount of ground mineral or other phosphate is scattered around each plant, normally about 2 ounces (*c.* 60 g). In most districts forestry

Table 3.1: Various cost details of afforestation for softwood, Sitka spruce, on 1 hectare of land in the UK. Figures with asterisks are used in calculations

Cost items	Years	Total cost, £, per hectare per year 1978/79 prices		
		Low	Average	High
(1) Ground preparation	0			
(a) Scrub cutting (by machine)		74	437*	800
(b) Draining		49	75*	101
(c) Ploughing (medium depth)		74	94*	114
(2) Planting	1		49*	
(3) Fencing (rabbit and sheep, approximately 120 metres)	1		495*	
(4) Weeding	2–3			
(a) By hand		47	60	72
(b) Chemically		30	36	42
(c) By machine		18	24*	30
(5) Brashing	10	62	82*	101
(6) Cleaning and pruning	15		59*	
(7) Road construction	15	50	100*	150
(8) Maintenance (drains and fences)	2–50		15*	

Sources: Forestry Commission (1963b, c, 1972, 1975b, 1978, 1979) and Hart (1979).

authorities have their own nurseries. Alternatively plants can be bought from commercial nurseries which can operate cheaply and more efficiently because of their large-scale operation. Planting-out age for Sitka spruce is normally 1–2 years. The cost of planting on a hectare of prepared soil is shown in Table 3.1.

Fencing

After young trees are planted the area must be fenced off against animals. There are various types of fencing to provide protection against sheep, rabbits, stock or deer. The fences are made up from a variety of woodwork ranging from squared oak to rough unsaleable softwood, in conjunction with steel wire, mesh or wooden boards. The Forestry Commission does not normally recommend the use of barbed wire unless local circumstances demand it.

Although fencing is an efficient method of protecting the forest from animals, it is an expensive operation. Sometimes natural boundaries may form an effective barrier to exclude animals and thus there may be no need for fencing. In some cases a fence may be obligatory because of some agreement in a lease or statute. Sometimes in a particular locality the damage which animals can inflict may be minimal. The probable amount of damage that animals may cause must be compared with the cost of fencing. For the types and methods of fencing see Forestry Commission (1972).

The larger the area fenced, the lower the cost per hectare. The cost per hectare of fencing areas from 4 to 40 hectares decreases considerably with increasing size. For areas over 40 hectares the decrease is negligible. In this analysis 120 metres of rabbit and sheep fencing per hectare is assumed. This type of fence includes mesh and wire supported by stakes. The analyst who is dealing with a specific area with specific problems has to carry out his own estimate. Table 3.1 shows the cost of fencing in various situations.

Weeding

Weeds frequently compete with newly planted tree crops during the early years of a plantation's life. The main risk to the crop arises from smothering and excessive shading. Smothering also creates conditions in which fungi can flourish. Although there are substantial variations in weeds and their vigour from site to site, some common types are bracken, bramble, coppice shoots and types of heather. The first three thrive on fertile soils and the latter are more common on infertile upland soils. Weeding on fertile soils can be an expensive operation as weeds are most persistent. Bracken grows quite quickly and overshadows the young plantation, and thus many foresters prefer to cut it twice a year.

Sitka spruce can force its way through lighter weeds and even bracken, which is sufficient to destroy most other trees. However, this species is more vulnerable to competition from heather than any other type of tree. A dense cover of heather causes nitrogen and phosphate deficiency, which is not easily corrected until the heather is controlled. This deficiency can be so severe that the crop almost ceases growing. Weeding can be done by hand as well as by chemicals, and its cost varies within wide limits and is dependent on the type of weed, the location, the methods of work and the skill of the staff. For full details of weed control see Forestry Commission (1975a). Table 3.1 shows costs of various methods of weeding on different soils.

Brashing

This means knocking off the lower branches of the crop trees. A variety of tools are available, such as pruning chisels, saws, secateurs, etc. It takes place about 10 years after planting and well before the first thinning. Before that a plantation requires little attention. When the trees are sufficiently tall to fend for themselves against competing weeds, they can be left alone until brashing. The reason for brashing is to provide easy access into the forest so that observation paths can be created to monitor the progress of tree growth. It also makes it very convenient for the thinning workers to work in the plantation. Indeed, pushing through in an unbrashed Sitka spruce plantation is an extremely difficult task. Brashing is usually done to the height of around 2 metres (about 6 feet), or to the nearest whorl below this height. This gives sufficient headroom for passage through the crop. Sometimes thinnings are to be sold standing, and very few merchants would be prepared to make an offer for unbrashed wood as they would normally want to see the merchandise beforehand. A costing for brashing is shown in Table 3.1.

Cleaning and pruning

Cleaning is the removal of weed growth between the trees. It can be done together with pruning. Weeds can be suppressed by the canopy of tree branches coming from the pruning operation. Some of these sticks may be commercially valuable as pea or bean sticks and firewood, but they are not always worth saving.

Pruning is carried out to secure knot-free clear timber and is usually done on the fast-growing conifers such as Sitka spruce. It is normally confined to the final crop of trees, i.e. the trees that will be kept until clear felling. Pruning is a way of investing in a tree. Unfortunately some pruned trees never reach full maturity. Some of them are blown down or, having lost their pre-eminence, are taken out in thinnings. The cost of pruning and cleaning for a typical case is shown in Table 3.1.

Road construction

Forests cannot be managed without roads. There is no other form of cultivation that is as dependent on roads as forestry. Timber is a very bulky and heavy commodity and in most cases it is produced in remote places away from a main road or railway. The roads that must be built inside the forest estate are more important for thinning than for clear felling. In the latter case it is possible to use heavy tackle and cut gradually into the plantation to winch the logs over long distances to points where they can be loaded on lorries. For thinning, however, the operation must be carried out between very long rows of trees without causing damage to the crop, and for this the inner roads are absolutely essential. Since thinning has to be carried out at regular intervals, then those roads will be in frequent use.

Because of the high cost of making roads, in most areas inferior tracks are established at a minimum cost to allow lorry transport, though only in reasonably dry weather. Good tarred roads can only be economic in very large plantations where they are used very frequently without restrictions on weight and speed. These roads also increase the value of standing trees and the land for future forestry operation. In the early periods of forest management in the British Isles, forest roads were constructed when the plantations were newly established. However, authorities soon realised that postponing road construction would improve the net present value figures because of discounting. Nowadays it is a part of good forestry management to delay road building as long as possible, usually until about year 15 just before the first thinning operation gets under way. This also shortens the period for maintenance. For details of optimal road spacing see Forestry Commission (1976).

Just as with fencing on a topographically homogeneous locality, the larger the size of the forest estate the lower the cost of road building per hectare. The problems of small plantations are those of low volumes of timber per cut, which makes harvesting more expensive because of a higher cost of fetching in equipment per cubic metre handled. The cost of any road into such forests would be prohibitively high. In recent years the development of high-capacity forwarders allowed timber extraction over agricultural land at relatively low cost with the minimum of damage to fields. This eliminated the necessity for building roads to outlying blocks of woodland. In this analysis a road construction cost range of £50 to £150 per hectare is assumed (Table 3.1).

Maintenance

In this, the most relevant items are maintaining fences, drains and roads and ensuring good hygiene to prevent damage by pests. Common insects in Sitka spruce plantations are the Pine Shoot Tartrix Moth, which feeds on the leading buds and causes distorted growth, the Pine Weevil, which causes damage to young plants by eating the bark, and *Adelges abeitis*, which causes damage by sucking plant juices. As regards the fungus pests, honey fungus is a common cause of death in young conifers.

Apart from all these costs there are other items which may become relevant in certain cases. These are beating-up, housing and fire protection. Beating-up, which is also called supplying or beating, is the replacement of plants that succumbed in the initial planting. In many plantations the take is so good that no beating is necessary. In a minority of cases the area has to be entirely replanted. Sometimes beating-up is required only in patches.

In remote plantations well away from towns and villages the forestry authority in charge may have to supply housing to forestry workers and administrators. In such cases housing cost should be included as part of the establishment cost. If there is more than one rotation, housing together with roads should normally be considered as part of the capital asset and be made available for several rotations.

Forest fires in the British Isles occur very seldom thanks to the wet and humid climate. Dry spells do not last long even during the high summer, and daily temperatures seldom climb over 25°C. The south east of England is drier and warmer in summer than any other area. Also this part of the country is densely populated, and frequent congestion occurs during the summer months in areas of scenic beauty where forests are located. This increases the fire risk.

Common causes of fires are uncontrolled moor burning by sheep farmers anxious to provide young heather growth for grazing, heather burning in banks of bogs before turf cutting, picnic fires, discarded cigarette ends and malice. Forest authorities maintain fire brigades which are on special alert during the danger periods. They also make use of the official local brigades and sometimes of the army. In certain locations there are fire look-out towers equipped with telephones. Unlike some southern European countries and the United States of America the fire-prevention cost in the British Isles is not very important in forestry. The damp climate of these islands is indeed a blessing for investors in forestry.

In this cost–benefit analysis no allowances are made for beating-up, housing and fire protection because they are either uncommon or expenses are too trivial to justify a special category of cost. Nor has any allowance been made to capture the administrative cost incurred by the Forestry Commission as a whole because a small acquisition does not add any significant amount to this cost. In other words, the administrative cost related to a marginal expansion is negligible. Of course, if the Forestry Commission were to include a large territory in its afforestation programme, then this would necessitate a substantial increase in the overall administrative cost which should be included in calculations.

Table 3.1 shows cost details for a typical Sitka spruce plantation in Britain, yield class 20, calculated on a per hectare basis. Obviously these costs vary considerably from place to place depending on different conditions and methods of work. However, an average level of expenditure can be estimated from these figures. In this respect figures marked with an asterisk, which are mostly the arithmetic average of high and low numbers, are employed in calculations. The gestation period of a single rotation is assumed to be 50 years, and 1978/79 base year prices are used throughout. The second column shows the years in which these costs are incurred.

The purpose of this exercise is to give the reader some idea about costs for the typical case. When the analyst is dealing with a specific plantation, which may not be comparable with the typical case, then he or she should make modifications in cost items. For example, he or she may be required to estimate the cost details of a plantation located on well drained and clean soil, a case in which scrub cutting and draining would be irrelevant.

Opportunity cost of land

This is defined as the foregone benefits which would have resulted had the land been employed in the next best alternative sector. Any given piece of land may yield a variety of products. It could be used to produce timber, beef, dairy or cereals, or could even be utilised for leisure purposes and urban development. The next best alternative to forestry is one that determines this cost. In most cases there is competition between agriculture and forestry for the utilisation of land. For a comprehensive study see the report of the Land Use Study Group (1966).

The forests in the UK can be grouped into three main categories. The first comprises small scattered forests surrounded by agricultural land, especially in areas too steep for other forms of cultivation. Most of the woodlands in the southern and midland counties of England belong to this group. A good proportion of them are in remote areas which can only be approached across fields and are therefore difficult to work. These areas are relegated to forestry because farmers do not want them. There is one advantage for these woods which may be set against numerous drawbacks. They have, in general, a much better quality of soil than is found in larger forest areas, a soil which may be good even by agricultural standards. On it trees grow fast and reach timber size on a relatively short rotation. The alternative land use in these areas may be dairy, beef or cereals.

The second group is composed of larger forests located on poor soils that have not been cleared for agriculture. Trees on these soils grow relatively slowly, and such areas were formerly preserved as royal forests and private chases. Some contain deep sands which are too unretentive for profitable farming but are good for trees. Forests in Dorset, Hampshire, Sussex and Wiltshire fall into this category. Woodlands in the heavy clay soils of the Midlands and South Downs can also be placed in this group. These forests are easy to reach, administer and work.

The third group of forests can be found in the mountainous areas and infertile soils of the United Kingdom, which are also known as less favoured regions. Forests in these areas are planted by the forestry authorities as well as by private foresters. Such locations are not suitable for profitable agriculture. Sheep farming appears to be the main alternative use for this category of land. In some cases land is not good enough even for sheep farming and it is left unused. Of course, the opportunity cost for the unused land is zero. A part of this land is also sub-marginal for forestry, which can yield a profit only when the price of timber becomes high. In other words, the limits of forest cultivation, like the limits of farm cultivation, may shift through changes in wood prices. They may also shift through improved methods of forest management and the introduction of new species of trees. However, it should be mentioned that climate and soil quality are not the only factors that affect the attractiveness of a forest site. Accessibility, proximity to market and the size of woodlands are also important factors in determining the value of land for forestry. Any forester who is seeking a site for plantation should take into account all these factors.

Most of the Forestry Commission's plantations are located in the third group. What is the opportunity cost of land in forestry which is established in less favoured locations? A survey carried out by the University of Exeter Agricultural Economics Unit revealed that in a typical less favoured area in England farming activities were divided between 44 per cent sheep farming, 46 per cent cattle farming, and the remaining 10 per cent other activities such as the growing of straw, cereals, etc. (Farm Management Survey, 1980). Table 3.2 gives output and cost details for a typical farm located on this type of land. In this, taxation and subsidies are excluded from the calculations in order to obtain more realistic figures. The net income after deduction of the farmer's labour cost and that of his wife turns out to be £66 per hectare. This figure is taken to be the opportunity cost of land throughout the life of the forestry project.

Table 3.2: Opportunity cost of land on a typical less favoured area, 1978/79 prices, per hectare

Total farm output (£)		272
Variable costs:		
Feedingstuffs	53	
Fertilisers	22	
Other costs	15	
Fixed costs:		
Labour (farmer, wife and other)	65	
Machinery	33	
Others (land maintenance, etc.)	18	
Total cost		206
Net income (opportunity cost or land rent)		66

Sources: Farm Management Survey (1980); Midland Bank (1980).

TIMBER BENEFITS

Timber benefits of forestry consist of two items: yield from thinning and yield from clear felling.

Thinning

Thinning means the cutting out of select trees from a plantation to gain a harvest and improve the growth and quality of the remainder. Economically, thinning can only be justified if the increased benefit from thinning outweighs the cost of it. There may be circumstances in which thinning may not yield benefits large enough to offset costs. In areas where road building is expensive, it may be worthwhile to postpone the construction of roads until the clear felling. Roads are built early mainly because of thinning operations. Also the cost of maintenance following thinning may be high. For example, expenditure on drain cleaning after thinning may be quite high, which would make the thinning worthless. Another case may be that the revenue difference between thinned and unthinned crop may be too small and it may pay off financially not to thin at all.

The major benefits of thinning lie in providing earlier returns in encouraging growth of the best trees so that the final yield becomes larger and higher in quality and also that it occurs earlier. Once a plantation is established, its well-being is largely determined by the skill with which it is thinned. Thinnings in the United Kingdom were formerly carried out in a mechanical fashion in which every third or fourth line of trees was removed. It saved marking and provided access for extraction. Today, foresters prescribe thinnings based on scientific research, and this yields better results. For a comprehensive discussion on various methods of thinning see Hiley (1967).

The Forestry Commission recommends a thinning regime governed by three factors.

(1) Intensity of thinning, i.e. the rate at which volume is removed from a stand of trees. The higher the thinning intensity the higher the space created for the main-crop trees. If these trees respond very well to the space created by way of high increases in mean diameter, then high-intensity thinning makes sense. In addition to this, higher thinnings also yield high thinning revenue. The maximum intensity that can be maintained without loss of volume production is called the marginal thinning intensity.
(2) Thinning cycle, i.e. the interval between successive thinnings. Long cycles result in heavier single thinnings whereas more frequent thinnings entail lighter single operations. The choice of thinning cycle depends on the yield class of the crop and on local management considerations. The Forestry Commission recommends a five-year cycle starting from year 15 for many yield classes of Sitka spruce.
(3) Thinning type. This refers to the type of tree that is to be removed in thinning. The main objective is to retain a sufficient number of trees to maintain vigorous growth on the site. One type is known as stem thinning in which

smaller trees are taken out first, leaving larger trees in a fairly even spacing. Another type is called crown thinning in which a number of the best, dominant trees are selected in a stand for the final crop and the spaces between them are occupied by smaller trees. In this the canopy becomes irregular and in some cases the stand appears to be unevenly aged.

Table 3.3 contains output details resulting from thinning. The figures are obtained from the Forest Management Tables produced by the Forestry Commission. Thinning starts in year 15 and continues every five years until the end of year 50. Yields from early thinnings are small but increase as the plantation becomes more mature.

Table 3.3: Yield (m^3) from thinning and felling, Sitka spruce, Class 20, 1 hectare. Cmtd means centimetre top diameter

Yield classes	Years							
	Thinning							Felling
	15	20	25	30	35	40	45	50
Cmtd around 7 (pulpwood)	13	68	64	52	32	19	12	90
Cmtd around 18 (palletwood and other products)	0	2	6	17	29	29	21	77
Cmtd over 24 (saw-wood)	0	0	0	1	9	22	33	404
Total volume, m^3	13	70	70	70	70	70	66	571

Source: Forestry Commission (1963a).

Felling

There are two types of felling: selective and clear. Some woods may contain unevenly aged trees, which may come about by design or by accident. In this, felling is normally confined to small groups, which gives better results than complete clearance all at once. Clear felling indicates a complete clearance of the site at the intended period. The method employed here is clear felling, which is assumed to take place at the end of year 50.

Felling is carried out by skilled and experienced foresters. Fellers are normally paid on a piece rate, and a skilled one can work two or three times faster than an ordinary woodman. When selective felling takes place, it requires more work and skill than clear felling as fellers have to climb trees and lop off large branches so that no harm can come to the young trees nearby. Normally conifers occupy less time per cubic metre to fell than hardwoods.

In forest management one does not have to stick rigidly to the intended period of rotation. Market conditions may change, favouring an earlier or later felling than was originally intended. Flexibility allows managers to take advantage of changing market conditions. A number of types of rotation are recognised by most foresters, namely technical rotation, rotation for maximum volume production, rotation for natural regeneration, and financial rotation. In the first the felling takes place at the time when the plantation meets the specification of a given market, e.g. veneer logs, saw logs, pulpwood, etc. In the second the objective is to determine a rotation which yields the maximum mean annual increment. Rotation for natural regeneration is determined by the age at which maximum viable seed is produced. Financial rotation is the one that yields a maximum discounted revenue.

The last column of Table 3.3 contains the yield resulting from clear felling at the end of year 50. The output is expressed in three classes as cmtd (centimetre top diameter) around 7, around 18, and over 24, which are known as pulpwood, palletwood and saw-wood. The bottom row shows the output resulting from thinnings at 5-yearly intervals, starting at year 15 and ending at year 45, with felling at the end of year 50. The felling takes place at an age which coincides with the maximum mean annual volume increment as seen in Figure 3.3. The volume of output from thinnings as well as from felling, and the age of maximum mean annual volume increment, change with the yield class. For example, for the yield class 8 the felling takes place around year 65. The output details for different yield classes of Sitka spruce and other species can be found in Forestry Commission Booklet No. 34 (1971).

PRICE OF TIMBER

As with the choice of discount rate, the price of timber is absolutely crucial in economic analysis of forestry investments. In view of the fact that forestry projects have long gestation periods there is a compelling need to estimate timber prices for the distant future. This is indeed a very difficult task. No investment project but forestry calls for economic analysts to predict prices that will rule in 50 or 120 years, depending on the species. During the lifetime of forestry projects many variables which determine wood prices may change quite dramatically. This will, no doubt, affect the fortune of investors in this field. The United Kingdom meets about 90 per cent of her wood requirement by imports. Therefore prices in the United Kingdom are dominated by world prices. In effect, the United Kingdom is the largest importer of wood in Europe. Table 3.4 shows some principal wood products importing and exporting nations.

In the world market the price of timber, just like the price of any other commodity, is determined by the forces of demand and supply. The cost of transport also plays an important role in determining local prices because wood is a bulky commodity in relation to its value. This gives protection to local growers in the United Kingdom. Transport by sea is much cheaper than by road, and no part of the British Isles is far from the coast. Therefore, timber growers in these islands are particularly exposed to foreign competition. Most exporting countries rely on virgin forests as opposed to plantations, a case which gives them a substantial cost advantage over the British growers who rely exclusively on plantations. As long as large areas of virgin forests remain, within easy reach of water transport, the price of timber in the British Isles will be determined mainly by the cost

Table 3.4: Major traders in wood products

Values of forest products, $ 000 000 1984	Canada	US	Sweden	Finland	West Germany	USSR	France	UK	Japan
Export	11 519	5 742	5 150	4 642	2 465	2 609	1 380	667	715
Import	1 003	10 618	499	337	5 091	959	2 534	5 329	6 131

Source: FAO (1984b).

of extracting, sawing and transport by the exporting countries. If the depletion of the world's forests continues without a fall in the rate of consumption, then it is likely that future prices will be much higher than they are at the present time.

The forests in the Third World are coming under increasing pressure. Lanly (1979) estimates that in the last 25 years about 136 million hectares of tropical forest have been destroyed. The situation in other parts of the developing world is even worse with the exception of India and China, where the forested area has increased slightly over the last couple of decades. The area under forest has shown a decreasing trend in the great majority of Third World countries. Table 3.5 shows the declining trend during the last 20 years. There are three main reasons for forest destruction.

(1) To meet the nutritional needs of the growing population, more and more forests have been converted into cropping lands: such deforestation is most extensive in Africa and Latin America.
(2) Commercially useful forests of many Third World countries are being cut down to meet the increasing domestic as well as world demand. During periods of economic decline some Third World countries eat into their forests to supply the world market with timber in order to earn precious hard currency (Kumar, 1982). Also, in their urge to generate employment, new paper and pulp industries have been started up without a long-term policy to protect the forest resources (Ahmad and Laarman, 1978).
(3) Woodlands are also being destroyed for firewood; this is a serious problem in Africa.

Simonis (1984) argues that one of the main reasons for the expansion of desert in Africa is the destruction of trees and other vegetation for firewood. The efficiency of fuelwood use in

Table 3.5: Decline of world's forests and woodlands

Year	1965	1970	1975	1980	1985
Area under trees, million ha	4200	4170	4159	4111	4030

Source: FAO (1960–85a).

Table 3.6: Growth of fuel wood production

Region	Quantity (000 m³)		
	1977	1980	1984
Africa	3 969	4 614	5 013
North America	12 358	33 598	33 418
South America	11 005	12 927	14 119
Asia	72 634	85 748	84 186
Europe	12 866	13 218	14 632
Oceania	125	305	625
USSR	54 300	51 300	52 200
World total	167 257	201 710	204 193

Source: FAO (1973–84a).

the Third World is extremely low. This is because many villagers use open cooking or inefficient wood-burning stoves. The alternative oil- or coal-based cookers are expensive and are beyond the means of the rural poor. Table 3.6 shows the growing trend in fuelwood production in the world. As oil and coal become more expensive this will make wood an even more valuable source of energy in the fuel-deficient Third World countries.

What is the likely scenario of timber prices in the distant future? It is obvious from the above arguments that the world's timber consumption is on the increase. On the other hand, the world's stock of timber resources is declining. Foresters have long realised that unless drastic steps are taken to stop the process of deforestation a timber famine will ultimately occur. Therefore, any sensible price projection for timber must be set against this background of increasing world demand and contracting stocks.

In 1962 two research associates, Potter and Christy, working for Resources for the Future Inc., an American non-profit making research establishment, analysed the consumption and price trends of many natural-resource-based commodities including pulpwood and lumber in the US between 1870 and 1957. Table 3.7 shows the long-term price trends of these two commodities. Whereas the price of pulpwood remained fairly stable, the price of lumber showed a very strong upward trend, becoming five times higher in 1957 than in 1870, an average annual increase of 1.9 per cent. Potter and Christy also noted that the rise in the price of lumber might be understated because of the shift of the industry from the eastern United States to the

Table 3.7: The long-term price trend of lumber and pulpwood in the US between 1870 and 1957, index 1957 = 100

Years	1870	1880	1890	1900	1910	1920	1930	1940	1950	1957
Real price of lumber	20	23	27	30	34	51	48	61	103	95
Real price of pulpwood	—	—	—	94	68	110	70	100	92	100

Source: Potter and Christy (1962).

west coast, which placed the output further away from the consumer market. Their study covered most principal commodities and lumber was the only one which showed a steep increase in price.

In the United Kingdom a number of statistics similar to those in Potter and Christy's work have been compiled to identify the long-term price trend of timber (see Hiley, 1930, 1967; MacGregor, 1953; Grayson 1967, 1974). A table summarising the situation in the United Kingdom over a hundred-year period between 1863/64 and 1963/64 is given in Hiley (1967). According to this, the annual average price increase of imported timber was 1.5 per cent in real terms. It showed a decline between 1834 and 1895 but then a strong upward trend prevailed until 1964.

What can we say about the price of coniferous timber during the lifetime of our project which has a 50-year gestation period? Accurate forecasting is an extremely difficult business but it is absolutely essential in this analysis. Five alternative assumptions can be made about timber prices over the next 50 years.

Assumption 1: The real price will not change at all.
Assumption 2: The real price will fall at an annual rate of 1 per cent.
Assumption 3: The real price will increase at an annual rate of 1 per cent.
Assumption 4: The real price will increase at an annual rate of 1.5 per cent.
Assumption 5: The real price will increase at an annual rate of 2 per cent.

Another important point is the choice of a base year. If 1952 or 1974 were chosen as a base year when the prices were very high, then this would overestimate all future prices. Conversely, if a low-price year was taken as the base, this would have the opposite effect. In the choice of a base year, 1978/79 is employed because the price in these years is very close to the average price between 1950 and 1980. Table 3.8 gives the average prices of standing coniferous timber in England, Wales, Scotland and Northern Ireland.

The price paid for a particular size or quality of timber depends on the supply and demand in a particular locality. In certain areas and at certain times there may be a greater demand for small trees. At other times and in other places, the small trees may be unsaleable as there is only a demand for sawlogs. It is also possible to find a situation where a tree is too big for sawmills in the locality and the price may actually fall. In places where there is a booming horticultural and agricultural industry, there may be a large demand for small trees for rustic poles, bean sticks and fence posts.

Prices for homegrown timber are dominated by supplies of sawn timber from countries such as Canada, the Soviet Union, Finland, Norway and Sweden, which are available to end users throughout the British Isles. No one is going to pay more for home-grown material than he has to pay for imported timber. When world prices rise, then home-grown prices rise. Conversely, a slump in world prices will produce a drop in the home market.

Table 3.8: Average prices (£/m^3) of coniferous timber sold standing. Figures with asterisks are estimates

Regions	Average volume per tree over bark (m^3), 1978/79 average						
	Up to 0.074	Over 0.074 to 0.124	Over 0.124 to 0.174	Over 0.174 to 0.224	Over 0.224 to 0.274	Over 0.274 to 0.424	Over 0.424
England	4.07	5.18	6.76	7.92	9.65	11.13	14.08
Wales	3.26	3.78	5.38	6.98	7.79	8.80	14.56
Scotland	2.45	3.20	4.85	6.38	8.07	8.32	10.86
Northern Ireland	1.77	3.00*	3.18	5.75*	5.65	8.06*	11.87*
UK average	2.89	3.79	5.04	6.76	7.79	9.08	12.84

Sources: Forestry Commission Annual Reports, HMSO London, and Northern Ireland Department of Agriculture Annual Reports, Government of Northern Ireland, Belfast, various years.

The figures shown in Table 3.8 indicate that there is a direct relationship between the average size of tree sold and the price paid per cubic metre. It is possible to group these size categories into pulpwood, palletwood and saw-wood. According to the Forestry Commission's Forest Management Tables all material under 0.224 m^3 per tree falls into the pulpwood category. Material between 0.224 m^3 and 0.424 m^3 per tree is palletwood, and all over 0.424 m^3 per tree is in the saw-wood category. Then the average prices for these three groups of wood for the base year of 1978/79 become:

Pulpwood	:	£4.62
Palletwood	:	£8.44
Saw-wood	:	£12.84

Table 3.8 also reveals that prices vary according to the area of the United Kingdom. In Britain, Scottish prices were lower than Welsh or English prices. But the regional differences may not stay this way all the time. For example, in 1977/78, prices in Scotland were higher than in Wales. Of course, if one is dealing with a particular location, one should take the prices prevailing in that region, instead of the United Kingdom average. A further refinement may be that in obtaining the average prices for each category one may wish to take the weighted average within each category. For example, in arriving at a £4.62 price for pulpwood I simply took the arithmetic average of the figures £2.89, £3.79, £5.04 and £6.76. In 1978/79 the distribution of revenue between all these four sub-groups was 17 per cent, 37 per cent, 24 per cent and 22 per cent, respectively. A weighted average then yields a price of £4.59. However, in a general cost–benefit analysis, which is the case here, such a practice would be a venture into triviality, although in the case of a specific location the exercise may well be justified.

Table 3.9 shows the details of the prices that may be applicable during thinnings and clear felling. With assumption 1 the real price of timber remains constant over time. With assumption 2, the figures are: for year 15 the price of pulpwood equals $£(4.62)(1-0.01)^{15} = £3.97$; for year 20 it is $£(4.62)(1-0.01)^{20} = £3.78$. With assumption 5 the price in year 15 is $£(4.62)(1+0.02)^{15}$ or £6.22; in year 50 it is $£(4.62)(1+0.02)^{50}$ or £12.44, and so on.

Table 3.9: Future price of wood (in £/m^3) in the UK under different assumptions

Assumptions regarding the future price of timber in the UK	Years							
	15	20	25	30	35	40	45	50
Assumption 1 (no change)								
Pulpwood	4.62	4.62	4.62	4.62	4.62	4.62	4.62	4.62
Palletwood	8.44	8.44	8.44	8.44	8.44	8.44	8.44	8.44
Saw-wood	12.84	12.84	12.84	12.84	12.84	12.84	12.84	12.84
Assumption 2 (1% decrease)								
Pulpwood	3.97	3.78	3.59	3.42	3.25	3.09	2.94	2.80
Palletwood	7.26	6.90	6.56	6.24	5.94	5.65	5.37	5.11
Saw-wood	11.04	10.50	9.99	9.50	9.03	8.59	8.17	7.77
Assumption 3 (1% increase)								
Pulpwood	5.36	5.64	5.92	6.23	6.54	6.88	7.23	7.60
Palletwood	9.80	10.30	10.82	11.38	11.96	12.57	13.21	13.88
Saw-wood	14.91	15.67	16.47	17.31	18.19	19.12	20.09	21.12
Assumption 4 (1.5% increase)								
Pulpwood	5.78	6.22	6.70	7.22	7.78	8.38	9.03	9.73
Palletwood	10.55	11.37	12.25	13.19	14.21	15.31	16.49	17.77
Saw-wood	16.05	17.29	18.63	20.07	21.62	23.29	25.09	27.03
Assumption 5 (2% increase)								
Pulpwood	6.22	6.87	7.58	8.37	9.24	10.20	11.26	12.44
Palletwood	11.36	12.54	13.85	15.29	16.88	18.64	20.58	22.72
Saw-wood	17.28	19.08	21.07	23.26	25.68	28.35	31.30	34.56

BRINGING THE RESULTS TOGETHER

Now we have all the ingredients of a discounted cash flow (DCF) analysis; output details, prices, costs and discount rates. Using the discounted cash flows method with various discount rates, e.g. 10 per cent test rate of discount, 5 per cent required rate of return and 3 per cent forestry target rate, the following results are produced. These are obtained by using the cost figures shown in Tables 3.1 and 3.2, the yield figures in Table 3.3 and the prices mentioned above.

	DCF with 10% interest rate	DCF with 5% discount rate	DCF with 3% interest rate
Assumption 1	−1852	−1693	−941
Assumption 2	−1903	−2077	−1818
Assumption 3	−1753	−1086	402
Assumption 4	−1691	−677	1356
Assumption 5	−1658	−624	1698

With a 10 per cent discount rate this forestry project fails, even with an annual 2 per cent increase in timber prices over time. Figure 3.4 shows the behaviour of the discount factor curve at a 10 per cent interest rate. The curve gets very close to the horizontal axis, which means practically a zero discount factor at the time when large benefits arise from felling. Figure 3.5 on the other hand shows the net cash profile of this project with constant prices. The net benefit curve remains continuously below the horizontal axis until the time when the second thinning takes place. From then on there are bumps in the curve resulting from thinnings. The cost is at the highest level when the project gets established during the first few years. With the exception of years 10 and 15 the annual cost between years 4 and 50 is constant at £81, which is the figure for the opportunity cost of land and maintenance. Discounting does have a devastating effect on early costs or benefits. For example, discount factors for the costs arising in years 0 and 1 are 1 and 0.9091, respectively. On the other hand the power of discounting at an interest rate as high as 10 per cent virtually wipes away the distant benefits that arise from late thinnings and felling. For example, the volume of sales at the end of year 50 from felling

Figure 3.4: Discount factor curve at a 10 per cent interest rate

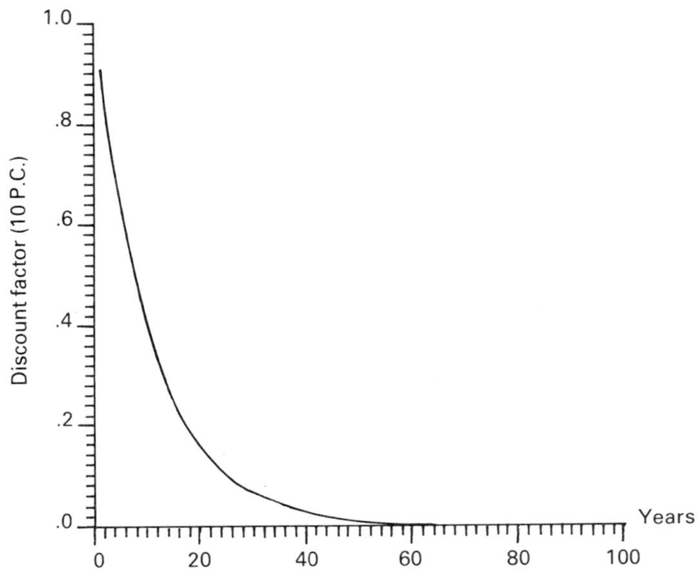

is £6253. But the discount factor for that time becomes 0.0085, which is practically zero and the discounted value of this sale becomes only £53. Table 3.10 shows the annual discounted as well as undiscounted net benefits, i.e. benefits minus costs, of this project with constant prices over time.

With constant and 1% annual reduction in timber prices the project fails to qualify even at a 3 per cent discount rate. However, the assumption of declining prices is unrealistic as there is no evidence to support it. Reductions in timber prices occur only temporarily and they are not a long-lasting phenomenon. As Figure 3.4 indicates, a decline is interrupted by a sudden upsurge in prices. Assumption 3 envisages a modest 1 per cent annual increase in real terms over time. Even with this the project does not qualify as a viable venture by using 10 and 5 per cent interest rates. With a 1.5 per cent annual price rise, assumption 4, the project becomes viable only at a 3 per cent interest rate.

Which is the likely scenario? It is essential to remember that the long-term price trend of timber is upwards. According to Hiley (1967), in the United Kingdom the annual price increase during the hundred-year period between 1864 and 1964 was 1.5 per cent in real terms. Some foresters believe that in view of the rapid decrease in world stocks this is a conservative figure for future prices.

Figure 3.5: Net annual cash stream of 1-hectare plantation Sitka spruce, yield class 20, with constant timber prices

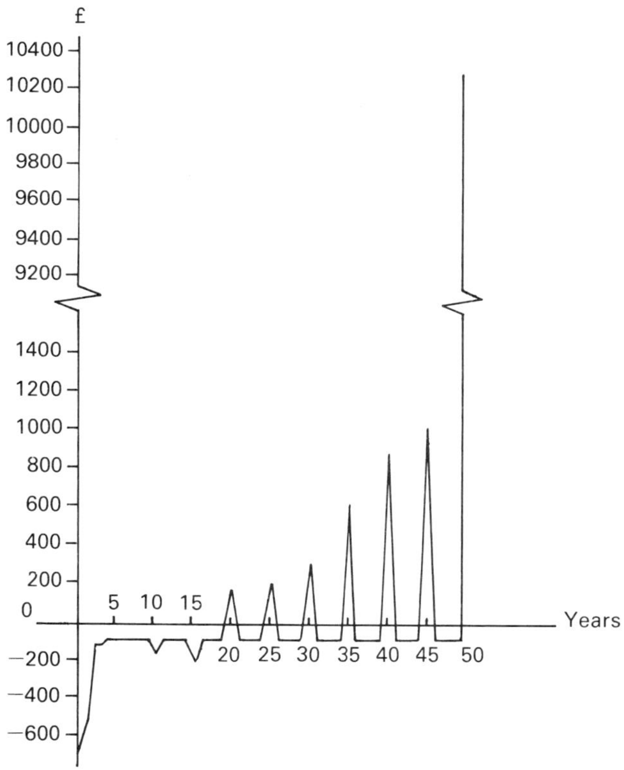

Table 3.10: Discounted (at a 10% interest rate) and undiscounted annual net benefits

Years	Net benefits undiscounted £	Net benefits discounted 10% £	Years	Net benefits undiscounted £	Net benefits discounted 10% £
0	−672	−672	26	−81	−6
1	−610	−555	27	−81	−6
2	−105	−87	28	−81	−6
3	−105	−79	29	−81	−5
4	−81	−55	30	316	18
5	−81	−50	31	−81	−4
6	−81	−45	32	−81	−4
7	−81	−41	33	−81	−4
8	−81	−38	34	−81	−4
9	−81	−34	35	427	15
10	−163	−64	36	−81	−2
11	−81	−28	37	−81	−2
12	−81	−26	38	−81	−2
13	−81	−23	39	−81	−2
14	−81	−21	40	534	12
15	−180	−43	41	−81	−2
16	−81	−17	42	−81	−2
17	−81	−16	43	−81	−2
18	−81	−15	44	−81	−2
19	−81	−13	45	575	8
20	250	37	46	−81	−1
21	−81	−11	47	−81	−1
22	−81	−10	48	−81	−1
23	−81	−9	49	−81	−1
24	−81	−8	50	6172	53
25	265	24			

4

Modified Discounting and its Application to Forestry

In this chapter the forestry project considered in Chapter 3 will be evaluated using the modified discounting method, which is also called the sum of discounted consumption flows. In this method, in view of the intergenerational distribution aspect of public sector investment projects, the conventional rules are altered yielding a much more moderate discounting pattern than the one in the net present value criterion. Forestry is an excellent example of the way in which many public sector projects redistribute income between generations. Its long gestation periods make it obvious that there is more than one generation involved in this venture. When felling takes place now, the present generations reap the benefits of a plantation which was established by earlier generations a long time ago. When planting takes place now, the benefits that will arise from it will be captured mostly by future generations. This phenomenon, however, is not confined to forestry. Almost all public sector investment projects do this to varying degrees depending upon the gestation periods involved. Evaluating these projects by means of the net present value or the discounted cash flows criteria results in discrimination against future generations. Let us prove how this happens. The following argument is extracted from the author's publications in the journal *Environment and Planning A*. Here it is presented in a modified form in order to provide easy reading. For full details of these publications see Kula (1981, 1984b and 1987a) and Price (1984b, 1987).

The net present value criterion is the most popular project evaluation method which is widely used in cost–benefit analysis. Despite its general popularity this method has limitations when it comes to evaluating long-term projects, particularly those in which the costs and benefits are separated from each other by a long time interval, which is the feature of many public sector projects. The discount factor $1/(1+x)^t$ (where x is the social rate of discount and t is the time), which is used to deflate future costs and benefits, is a decreasing function of time and approaches zero over time. Multiplying distant costs and benefits by the relevant factor reduces them to insignificant figures. The higher the discount rate, the faster the discount factor approaches zero, reducing the present value of these future costs and benefits to practically nothing. Under the rules of the net present value criterion the distant consequences of the projects become immaterial and do not play a decisive role in the decision-making process. Some economists argue that the net present value criterion may be a reasonable means of comparing short-run projects, but for long-term projects in which a substantial redistribution of income is involved between generations it does not provide a satisfactory answer. In this respect Nash (1973) and Page (1977) argue that by following the rules of this method present members of society could pass great costs on to future generations by undertaking projects, the heavy costs of which would appear in the distant future, or fail to undertake projects which may have a very low construction and maintenance cost now but yield large benefits in the years to come. Below, with the help of a simple model, it is proved that the net present value method of investment evaluation discriminates against future generations in public sector economics.

FUTURE GENERATIONS AND THE NET PRESENT VALUE METHOD

The argument is advanced with the help of a simple illustrative example, and by making a number of simplifying assumptions:

(1) It is assumed that a society consists of only three individuals, namely person A, person B and person C.
(2) The life expectancy of the individuals in this hypothetical society is three years and the individuals are of different ages. Person A is the oldest member of the society being two years old and having one more year to live; person B is one year old and has two more full years to live; person

C is the new arrival and has three full years ahead of him.

(3) The size of the population and the life expectancy of its members are stable, that is, there are always three members and the life expectancy of each member is three years. Individuals are mortal, but the society lives on. In the first period there exist three individuals, A, B and C. At the beginning of the second period, person A dies and is replaced by a newcomer, person D. At the beginning of the third period person B dies and person E joins, and so on.

(4) In the first period with A, B and C, the society decides to undertake a project, the costs and benefits of which extend over many generations. The project has an economic life of, say, eight years and as a consequence, in addition to the decision-making individuals, future members of society who are yet to be born are affected.

(5) The initial construction cost of the project, say £300, is shared equally by the existing members of the society, that is, A, B and C. These costs, *C*, are regarded as negative benefits accruing to them, that is, $C_A = -£100$; $C_B = -£100$; $C_C = -£100$. It is also assumed that the money comes from the source displacing each individual's private consumption by that amount.

(6) From the second year onwards the project yields net consumption benefits over its lifetime. The certainty equivalent of these benefits is £300 every year. When these net benefits become available at the end of each period, they are distributed equally between the existing members of society who are alive at the time. Each beneficiary consumes his share of benefit immediately it becomes available. At the beginning of the ninth year the project is written off with no cost.

(7) At the beginning of the first period, the project is evaluated under the rules of the net present value criterion. One of the members, say person A, undertakes this job on behalf of the society. Since it is assumed that the project is financed through a reduction in individuals' private consumption, and the resulting benefits are also consumed by the individuals when they become available, the analyst needs to know the social time preference rate (also called the consumption rate of interest) to deflate the consumption stream and obtain a net present value for this project. It is also convenient to assume that all members of society, at any point in

time, have the same time preference rate of, say, 10 per cent. Then the social time preference rate would be 10 per cent at all times, a figure reflecting all individual rates and averaged over all individuals who are alive at the time.

The decision-maker is now permitted to evaluate the worth of the hypothetical project under the rules of the net present value criterion, which is

$$NPV = \sum_{t=1}^{8} \frac{1}{(1+s)^t} NB_t \tag{4.1}$$

where NB_t is the net benefit arising at time t; s is the social time preference rate, or the consumption rate of interest, which is 10% throughout, t is the time which goes from year 1 to the end of year 8.

Substituting numerical values gives:

$$NPV = -\frac{300}{(1.10)^1} + \frac{300}{(1.10)^2} + \cdots + \frac{300}{(1.10)^8}$$

$$NPV = 1056.11$$

Now in view of assumption 6, if equation 4.1 is rewritten in detail by taking into consideration who receives the benefits and pays the costs over time:

$$\begin{aligned} NPV = & \left[\frac{-100}{(1+s_A)} + \frac{-100}{(1+s_B)} + \frac{-100}{(1+s_C)}\right] \\ & + \left[\frac{100}{(1+s_B)^2} + \frac{100}{(1+s_C)^2} + \frac{100}{(1+s_D)^2}\right] \\ & + \left[\frac{100}{(1+s_C)^3} + \frac{100}{(1+s_D)^3} + \frac{100}{(1+s_E)^3}\right] \\ & + \cdots + \left[\frac{100}{(1+s_H)^8} + \frac{100}{(1+s_I)^8} + \frac{100}{(1+s_J)^8}\right] \end{aligned} \tag{4.2}$$

In equation (4.2) one simply takes into account births and deaths, who receives how much and when; s_i is the time preference rate of the ith person alive at that time who receives the benefits or pays the costs, as the case may be. Since by assumption 7 all individuals' time preference rates are 10 per cent, these rates are substituted into equation (4.2) to give the same numerical result, which is £1056.11.

Equation (4.2) which gives an identical result to equation (4.1), brings out a very important argument upon which is based the net present value criterion. In the first year, the members of society who actually pay for the project (persons A, B and C), wait one year to incur their share of cost. Their subjective valuation of the cost, seen from the date at which they became aware of it, is expressed in the first square bracket of equation (4.2). The worth of this cost to society, seen from the beginning of year one, is $(-1/1.10)300$.

In the second year, however, person A is no longer in the society. He died at the end of the first year, immediately after incurring his share of the cost, and is replaced by a newcomer, person D. The worth of the second round of benefits to the society seen from the beginning of the first year is expressed in the second square bracket. Persons B and C, indeed, wait a further year for the second round of benefits which they acquire at the end of year two. For person D, however, equation (4.2) assumes that he also waits for two years to get his share. In fact he does not: he waits only one year for his share.

In the third year the situation is the same. The decision-maker looks at the third round of benefits from the vantage point of the first year and assumes that each recipient waits three years for his share. Person C's share of £100 is deflated by the discount factor $1/(1+s_C)^3$ assuming that he waits for three years, which indeed he does. On the other hand, persons D and E wait for only two years and one year, respectively, to acquire their shares which arise at the end of the third year.

The same treatment is given to the individuals who receive their shares of the benefits at the end of the fourth ... eighth years, that is, it is assumed that they all wait for four ... eight years, respectively.

Table 4.1 summarises the relationship between the project and the members of society over time. Column 7 illustrates the number of years each beneficiary waits in order to acquire his share of benefits. Column 8 on the other hand shows the power

Table 4.1: The relationship between the project and the members of society over time

Period	Total net benefits arising	Who dies at the end of the period	Who is born at the beginning of the period	Who is alive at the time	Their share of net benefits	Number of years they waited to acquire their shares	Power of $(1 + s)$	Weights given to the share of each beneficiary
(1)	(2)	(3)	(4)	(5)	(6)	(7)	(8)	(9)
1	– £300	—	—	A	– £100	1	1	$1/(1.10)$
				B	– £100	1	1	$1/(1.10)$
				C	– £100	1	1	$1/(1.10)$
2	£300	A	D	B	£100	2	2	$1/(1.10)^2$
				C	£100	2	2	$1/(1.10)^2$
				D	£100	1	2	$1/(1.10)^2$
3	£300	B	E	C	£100	3	3	$1/(1.10)^3$
				D	£100	2	3	$1/(1.10)^3$
				E	£100	1	3	$1/(1.10)^3$
4	£300	C	F	D	£100	3	4	$1/(1.10)^4$
				E	£100	2	4	$1/(1.10)^4$
				F	£100	1	4	$1/(1.10)^4$
5	£300	D	G	E	£100	3	5	$1/(1.10)^5$
				F	£100	2	5	$1/(1.10)^5$
				G	£100	1	5	$1/(1.10)^5$
6	£300	E	H	F	£100	3	6	$1/(1.10)^6$
				G	£100	2	6	$1/(1.10)^6$
				H	£100	1	6	$1/(1.10)^6$
7	£300	F	I	G	£100	3	7	$1/(1.10)^7$
				H	£100	2	7	$1/(1.10)^7$
				I	£100	1	7	$1/(1.10)^7$
8	£300	G	J	H	£100	3	8	$1/(1.10)^8$
				I	£100	2	8	$1/(1.10)^8$
				J	£100	1	8	$1/(1.10)^8$

Source: Kula (1981).

to which the beneficiaries' discount factor rises in the net present value criterion to deflate the benefits which are accruing to them so that an overall net present value figure can be obtained.

In this argument, assumptions were made with regard to the population structure of the society, and this represents the real world situation in a simplified but realistic fashion. Assumptions 1 and 2 point out the fact that by birth and death the constituent members of the society continuously change over time. If each person is regarded as representing a different generation, assumptions 1 and 2 hold approximately true for societies in which there are, more or less, three generations in existence at a point in time with their lifespans overlapping. In assumption 3 it is argued that in the hypothetical society the size and age structure of the population is constant. The assumption is not unrealistic, at least as far as the experiences of industrialised European countries are concerned. Now, the question is what kind of assumptions permit the net present value criterion to treat in this way the future members of society, who are not alive at the time when the decision is made?

A possible explanation of the way in which the future benefits and costs are discounted in such a manner may be that it is implicitly assumed in the net present value criterion that the welfare of the future members of society, who are not present at the time the decision is made, is somewhat less important than the welfare of those who are alive at that particular time. Thus the benefits of the project which will be accruing to those individuals in the future are given less weight in comparison with those who are presently alive; and the more distant they are, the less weight is attached to their shares. For example, from Table 4.1 it can be seen that, in the NPV method, for the benefits arising in the seventh year, person I waits only one year for his share and is given a weight of $1/(1.1)^7$. For the benefits arising in the eighth year, person J, who is a more distant member than person I, also waits for one year to acquire his share, but this is given a weight of $1/(1.1)^8$.

How can such an assumption be justified — that is, the assumption that the welfare of future generations is less important than that of the present ones? Should the decision-maker in his analysis of a public investment project give less weight to the benefits that will be accrued by future members of society? Should person I's share of benefit, accruing to him at the end of the eighth year, be given less weight *vis-à-vis* those of persons B and C, accruing to them at the end of the second year, despite the fact that they all (B, C and I) wait for an identical length of time (two years) to acquire their shares?

This practice, that is, undervaluing future generations' shares in comparison with present ones, is in conflict with the views of a number of writers who have made explicit statements on this subject. Pigou (1929) argues that 'It is the clear duty of government, which is the trustee for unborn generations as well as for its present citizens ...'. Marglin (1963b) argues likewise: 'since the generations yet to be born are every bit as important as the present generations ...'. These authors indicate that, in their view, the welfare of each generation, present and future, is equally important, and thus it follows that government should give equal weight to each generation's share of benefits (or losses).

More recently, Rawls (1972) has developed a theory of social compacts as a basis for ethics. However, Rawls devotes most of his efforts to the intratemporal case with only passing reference to the intertemporal one. In his argument regarding intertemporal fairness and determination of just saving principles over time, Rawls treats each generation equally. In the original position, representatives from all generations are assumed to know the same facts and to be subjected to the same veil of ignorance, and also they will probably be given one vote to cast their opinion during the possible discussions with regard to choosing one of the many methods of carrying out the policies that are agreed on. In the original position it is hard to imagine that any method like the net present value criterion, which gives unequal weights to the receipts of different generations, will be accepted as fair. Apart from the question of fairness, it is unlikely that representatives will accept this evaluation method. This is because under the rules of this method it is possible for one generation to pass costs on to the succeeding ones. Since delegates would not know to which generation they belonged, they may be at the receiving end of the costs. In order to safeguard every interest, it is more likely that they would promote some other project evaluation methods giving equal weight and importance to the welfare of all generations. In the end, what is good for one would be good for everyone. It is not considered that in the original position the delegates would deny discounting. It was one of the conditions that in the original position the representatives should know the general facts of human society and laws of human psychology. Since individuals, in normal circumstances, discount their future utilities and disutilities and would not be indifferent between their preference of a present consumption benefit over a future one of the same magnitude, there would therefore be no grounds for delegates to disregard this preference of individuals. However, it should be mentioned that some economists such as Feldstein (1964) have argued that a zero time preference rate is possible, a situation in which individuals, at the margin, will be indifferent to a given amount of present consumption and a future one of the same magnitude. There is also evidence that, for much of the post-war inflationary period, individuals have been willing to save at negative real interest. However, assuming that discounting is the usual behavioural pattern of human beings, then in the 'original position' there will be a strong case for its acceptance provided that all representatives are equally treated, although each may have a different rate of time preference.

In view of the illustrative example, one can hardly claim that by using the net present value criterion the decision-maker has given equal weight to the shares of benefits accruing to future members of society, that is, persons D, E, F, G, H, I and J.

THE MODIFIED DISCOUNTING METHOD, MDM

In the hypothetical project, there is no quarrel with the net present value criterion as far as the treatment it gives persons A, B and C is concerned. The actual time that those individuals wait for their share of benefits and costs is properly taken into account in equation (4.2). If the same procedure is repeated for all individuals and the discounted net consumption benefits are summed, this gives an overall figure. This method is called modified discounting or the sum of discounted consumption flows, which measures the overall worth of a project to individuals who actually receive the net benefits arising from it.

The rules of this method are as follows. A public project provides a fixed and unalterable stream of consumption benefits to society, the constituent members of which continuously change over time. Since there is no market for the beneficiaries to exchange their share of the return from a public investment programme, they can only wait to acquire those net benefits. As

Table 4.2: The undiscounted and discounted net benefit shares of individuals

Person[a]	Time[b]	Undiscounted net benefits									Discounted net benefits								Discount consumption flows
		t_1	t_2	t_3	t_4	t_5	t_6	t_7	t_8	Total	t_1	t_2	t_3	t_4	t_5	t_6	t_7	t_8	
A	t_1	−100								−100	$-\frac{100}{(1.10)}$								−90·91
B	t_1	−100	100							0	$-\frac{100}{(1.10)}$	$\frac{100}{(1.10)^2}$							−8·27
C	t_1	−100	100	100						100	$-\frac{100}{(1.10)}$	$\frac{100}{(1\text{-}10)^2}$	$\frac{100}{(1.10)^3}$						66·92
D	t_2		100	100	100					300		$\frac{100}{(1.10)}$	$\frac{100}{(1.10)^2}$	$\frac{100}{(1.10)^3}$					248·74
E	t_3			100	100	100				300			$\frac{100}{(1.10)}$	$\frac{100}{(1.10)^2}$	$\frac{100}{(1.10)^3}$				248·74
F	t_4				100	100	100			300				$\frac{100}{(1.10)}$	$\frac{100}{(1.10)^2}$	$\frac{100}{(1.10)^3}$			248·74
G	t_5					100	100	100		300					$\frac{100}{(1.10)}$	$\frac{100}{(1.10)^2}$	$\frac{100}{(1.10)^3}$		248·74
H	t_6						100	100	100	300						$\frac{100}{(1.10)}$	$\frac{100}{(1.10)^2}$	$\frac{100}{(1.10)^3}$	248·74
I	t_7							100	100	200							$\frac{100}{(1.10)}$	$\frac{100}{(1.10)^2}$	173·55
J	t_8								100	100								$\frac{100}{(1.10)}$	90·91
Total		−300	300	300	300	300	300	300	300	1 800	−272·73	256·55	248·68	248·68	248·68	248·68	248·68	248·68	1475·90

[a] That is, the beneficiary.
[b] That is, the time when each beneficiary became aware of their future net benefits.
Source: Kula (1981).

regards the discussion concerning the characteristics of public projects that provide an unalterable stream of consumption returns to the society which cannot be exchanged or assigned away, see Broussalian (1966, 1971). Those individuals at some point in time become aware of the fact that some fixed future consumption benefits (or costs) are created for them by the project. Discounting these net consumption benefits by their subjective time preference rates, they obtain a net present value figure for themselves at that date. Each figure summarises the recipients' subjective evaluation of these net benefits which will accrue to them at some future date. In fact, this practice is done for persons A, B and C in the net present value criterion. In the example, the project creates a fixed stream of consumption benefits over time for many individuals, most of whom are yet to be born. Persons A, B and C, being alive at the precise moment of decision-making, become aware of the fact that they have benefits stored for them in the future (for person A it is the cost). By discounting at their subjective time preference rate, they bring those future benefits back to the time when they were made aware of what was lined up for them in the future. The same procedure is repeated for each beneficiary, and an overall figure is obtained by summation. This figure indicates the worth to society of the consumption stream created by the public project. If this figure is positive, the project is worth undertaking. Among the exclusive projects, the one with the highest sum will be the most preferred investment.

Table 4.2 shows the relationship between the hypothetical project and the individual members of society over time. The first column shows the recipients of the benefits arising from the investment project, and the second column shows when individuals become aware of their share of future benefits, which is at the beginning of each period, shown at $t_1 \ldots t_8$.

As far as the rules of the sum of discounted consumption flows are concerned, a cost–benefit analyst, if he is acting on behalf of all individuals (present as well as future), should estimate what proportion of the benefits will be received by each person and when. He should also estimate when each beneficiary will become aware of what is in store for him in the future and discount each recipient's benefit stream back to this precise point of time. Persons A, B and C become aware of their share of net benefits at t_1, person D at t_2, ... person J at t_8. The analyst discounts each individual member's net benefit streams as seen from these time points. The last column of the table shows the discounted benefits of each recipient; in this respect person A's discounted net consumption benefit is −£90.91, person B's −£8.27, ... person J's £90.91. These discounted shares of each recipient are summed, and the overall figure, £1475.90, is shown at the bottom right-hand corner of the table.

By means of the net present value formula a figure of £1056.11 is obtained, whereas the sum of discounted consumption flows gives £1475.90. The latter shows a much higher consumption value due to the fact that individuals who receive future benefits but are not alive at the precise time the decision is made are given exactly the same treatment as the ones who are alive.

The simple illustrative example in this chapter enables one to formulate the modified discounting method. In this respect two formulae can be written which give identical results. This way one has a means of checking that the results obtained are correct.

A formula based on the annual discounted sums

Table 4.2 indicates that the figure of £1475.90 can be reached either by summing the last row horizontally, or by summing the last column vertically. The figures in the last row are obtained by first summing vertically the annual discounted net benefits of the project and then by totalling those sums horizontally.

The first two columns are different from the others. This is because in the third year we reach a point where, from a discounting point of view, the relation between the project and the individuals becomes repetitive. Therefore, the summing-up process can be separated into two stages: summing until the beginning of the third year, and summing from the third year onwards.

The formula itself is:

$$\text{MDM} = \sum_{t=1}^{n-1} \left[\sum_{k=2}^{t} \frac{\text{NB}_t}{I^{(k-1)}} + (n+1-t)\frac{\text{NB}^t}{I^t} \right] + \sum_{t=n}^{N} \left[\sum_{k=1}^{n} \frac{\text{NB}_t}{I^k} \right] \tag{4.3}$$

where

NB_t is the net benefit of the project accruing to each individual at time t;
n is the life expectancy of individuals, which is three years in the example;
t is the age of the investment, which goes as $t = 1 \ldots 8$;
N is the life of the project, which is eight years in the example;
k is an index number for calculation representing each cohort recruited into the population; and
I equals $(1+s)$ where s is the social time preference rate, which is 0.10 in the example.

The first term on the right-hand side of the equation is related to the first two columns of the discounted net benefits in Table 4.2, and the second term is related to the remaining columns.

A formula based on each individual's total discounted share

In Table 4.2 the figure £1475.90 can also be obtained by summing the last column vertically, but first the column figures are obtained by summing each beneficiary's total discounted share horizontally.

As far as the denominators are concerned, the first two rows and the last two rows are different from the rest, and therefore the rows can be separated into three groups: first two rows, repetitive rows in the middle, and last two rows. The first two rows are related to the individuals who were born before the project commenced, the repetitive rows are related to the individuals who will be born after the start of the project and die before it ends, and the last two rows are related to the individuals who will be born after the start of the project but outlive it. This argument is summarised in the following formula:

$$\text{MDM} = \sum_{d=n+1}^{-1} \left[\sum_{h=1}^{n+d} \frac{B_h^{NT}}{I^h} \right] + \sum_{d=1}^{N} \left[\sum_{z=0}^{n-1} \frac{B_{d+z}^{\text{NT}}}{1^{z+1}} \right] \qquad (4.4)$$

subject to $B^{NT}_{d+z} = 0$ when $d+z \geqslant N$, and where

B^{NT} is the total net benefit accruing to each individual;
d is the beneficiary's birth date relative to the starting date of the project ($d = -1$ means that the project was undertaken when the beneficiary was one year old, $d = 1$ means that the beneficiary was born when the project was one year old);
h is an index number including only the beneficiaries who were born before the project started; and
z is the age of the individual in question.

The first term on the right-hand side of equation (4.4) is related to the first two rows, the second term is related to the repetitive rows, and the condition takes care of the last two rows in the discounted net benefits of Table 4.2.

There are some notable differences between the modified discounting method and the net present value criterion. These are:

(1) In the net present value method the net present value is a function of time and discount rate. On the other hand, the overall figure obtained by the modified discounting method is a function of life expectancy of individuals and size of population, as well as the rate of discount. Changes in these factors will have an impact on the final figure. However, equations (4.3) and (4.4.) will hold for the countries in which the size and structure of the population is more or less stable, which is the case for many industrialised European countries.
(2) In the net present value criterion there is a single time point, the present. The entire future economic consequences of the project are brought to that point, which is common for all investment projects under comparison. In the modified discounting method, however, there is not a single time point, but rather a number of points, each representing a period when an individual recipient becomes aware of his future share of benefit (or cost) which will accrue to him as a result of the project in question. These time points are common for all projects under comparison. Of course, projects which have longer economic life will affect a larger number of future members of society and will have more time points than projects which have shorter economic life.

(3) In the net present value criterion the discount factor diminishes with time. In this way, for long-term projects the distant economic consequences are reduced to insignificant figures and do not play an important part in decision-making. In other words, with a positive social rate of discount, the power of the discount factor practically wipes out the present value of the distant benefits, which are the feature of many public sector projects. On the other hand, in the modified discounting method the discount factor does not decrease continuously but remains constant from the point when the relationship between the project and the recipients is stabilised (this point in the example is period 3 — see the last row of Table 4.2). In this way the distant consequences of projects become important in a decision-making process. In the case of short-run projects, the results obtained by either method will not differ significantly. As regards long-living projects, the lifetimes of which extend over many generations, the modified discounting method will yield a figure considerably different from that obtained by the net present value criterion.

However, it should be pointed out that the modified discounting method brings its own problems. Since in this method one does not have a continuously declining discount factor approaching zero over time, the costs and benefits of a project which arise in the distant future become important. Undoubtedly, forecasting the relative prices for all future dates would be a very difficult practice, but this is a different argument altogether.

DEVELOPING A MODEL TO ESTIMATE THE MODIFIED DISCOUNTING METHOD

We are now in a position to set up a model to estimate discount factors for various interest rates. The modified discounting method will be influenced by a number of factors. These are: the magnitude of the discount rate, growth of population and the life expectancies of members of society. Obviously, these variables differ from one country to another. In the United Kingdom over the last 15 years the population has remained around the 56 million mark and there is no reason to believe that it will change substantially one way or the other in the foreseeable future. Therefore, it will be reasonable to assume a zero growth population in our model. With regard to the expected lifetime of individuals, it should be noted that the life expectancy of United Kingdom citizens has drifted upwards over the years. The figure for the life expectancy of females at birth has gone up from 75 years in 1966 to 76 years in 1981. For males it has gone up from 69 years to 70 years during the same period (*Annual Abstract of Statistics*, 1980). Although it is conceivable that these figures may continue to increase over the years, the statistics indicate that this is likely to happen only very slowly. For the time being a constant life-expectancy figure of 73 years as an average for both sexes is assumed in the United Kingdom. Later on a sensitivity test will be carried out to find out to what extent increasing life expectancies over time may alter our results. One further assumption is that the structure of a population consists of equal cohorts, and that recruitment into that population (necessary with the assumption of constant population) equals deaths.

The next step is to imagine a situation in which a public sector project yields £1 net benefit every year (in the case of net cost, it can be expressed as −£1) which is divided equally between the members of society who are alive at the time. This modifies equation (4.3) into figures which form the rows of a matrix that yields discount factors for various rates of discount over time. The rows of the matrix are formed by estimating the following terms as the discount varies from 1 to 15 per cent across each row:

$$\frac{n(1/n)}{I} = \frac{1}{I} \qquad \text{(row 1)}$$

$$\left[\sum_{k=2}^{t} \frac{(1/\text{n})}{I^{k-1}}\right] + \frac{(n+1-t)(1/n)}{I^{t}} \qquad \text{(row 2 to row } n-1\text{)}$$

$$\sum_{k=1}^{n} \frac{1/n}{I^{k}} \qquad \text{(row } n\text{)}$$

where

n is the life expectancy of individuals, which is 73 years; it also represents the single age groups into which the population is divided;

I equals $1+s$, where s is the discount rate, for which figures between 1 and 15 per cent are chosen;

t is the age of the investment which may go from one year to infinity. It also takes the value of the specific row concerned; for example, in row 2, $t=2$, whereas in row $(n-1)$, $t=(n-1)$;

k is a number which refers to the kth cohort recruited into the population.

On the basis of the above model a computer program, written in BASIC, is used to estimate the discount factor matrix for the United Kingdom. The same program can be used to estimate similar factors for different countries or for different discount rates. A discount factor matrix which contains discount factors for interest rates between 1 and 15 per cent is provided in Appendix II.

```
10   DIM A(200,15)
20   INPUT ''INPUT THE NUMBER OF ROWS YOU
     REQUIRE (N<=200) '';N
30   T=1
40   FOR I=1 TO 15
50   A(T,I)=1/(1+(I/100))
60   NEXT I
70   REM ''ROW 1 NOW FORMED''
80   REM ''T IS ROW, I IS COLUMN''
90   FOR T=2 TO N-1
100  FOR I=1 TO 15
110  S=0
120  FOR K=2 TO T
130  REM ''K IS THE COUNTER FOR SUMMATION''
140  S=S+(1/N)/(1+(I/100))^(K-1)
150  NEXT K
160  A(T,I)=S+(N+1-T)*(1/N)/((1+(I/100))^T)
170  NEXT I
180  REM ''ROW T FORMED''
190  NEXT T
200  REM ''ROWS 2 TO N-1 NOW FORMED''
210  FOR I=1 TO 15
220  S=0
```

```
230  FOR K=1 TO N
240  S=S+(1/N)/(1+(I/100))^K
250  NEXT K
260  A(N,I)=S
270  NEXT I
280  REM ''ROW N NOW FORMED''
290  REM ''NOW PRINT OUT MATRIX''
302  PRINT TAB(30);''DISCOUNT FACTOR''
304  PRINT  \ PRINT  TAB(31);''PERCENTAGE(I)''
306  PRINT  ''YEAR(T)'';
307  FOR I=1 TO 15
308  PRINT USING ''££'';I;
309  NEXT I \ PRINT
310  FOR T=1 TO N
315  PRINT  USING ''££'';T;
320  FOR I=1 TO 15
330  PRINT USING ''  .£££'';A(T,I);
340  NEXT I
350  PRINT
360  NEXT T
370  END
```

Figure 4.1, which is accurately drawn by computer, illustrates the difference between discount factors that belong to the net present value and the modified discounting criteria resulting from a 10 per cent discount rate. The space between the two curves, shown as the shaded area, may be called the discrimination gap, created by the discriminatory discounting method practised by the net present value criterion against the future members of society. This gap is narrow in the early years, but widens as time passes. If a public sector investment project were a short-lived one, the level of discrimination against future generations would be low when evaluated by means of the net present value criterion. For investments with a long life — a feature of most public sector projects — the net present value rule tends to practise a very high degree of discrimination against future members of society.

Unfortunately, in practice when the net present value criterion is used it is quite normal to take a cut-off period of 25 to 30 years and to ignore the costs and benefits of public projects beyond that point. In fact, a 30-year cut-off period was recommended by the Ministry of Overseas Development (1977) to

Figure 4.1: Discount factor curves produced by means of the net present value and the sum of discounted consumption flows criteria at a 10 per cent interest rate

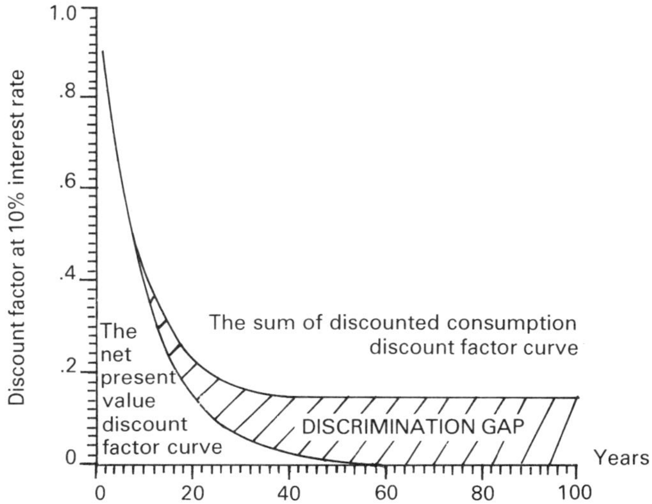

Source: Kula (1984b).

evaluate the worth of public sector investment projects overseas. In effect, the estimated lifetime of many public sector projects stretches well beyond this cut-off period. For example, single- and multi-purpose river projects, transport projects such as motorways, bridges, air and seaports, forestry projects, and many other social and environmental projects, all have life expectancies well over the 30-year cut-off period. Figure 4.1 makes it obvious why a cut-off period of this kind is adopted in practice when analysts evaluate the worth of public sector investment projects by means of the net present value criterion. It is a fact that, with a discount rate as high as 10 per cent, benefits and costs that occur beyond the immediate few decades get wiped out anyway and thus there is no point in giving them proper attention. The higher the rate of discount the shorter the time horizon beyond which costs and benefits become immaterial. If these projects are evaluated by means of the modified discounting method, then a different picture will emerge.

However, it is also obvious from Figure 4.1 that the two discount factor curves run fairly close together in the early years and therefore the ordinary rule may be used (approximately) for short projects, say up to ten to twelve years, without substantially violating the modified discount rule.

As for the assumption regarding a constant 73-year average life expectancy for United Kingdom citizens, a sensitivity test has been carried out by envisaging a situation in which the average life expectancy drifts upwards from 73 years towards 79 years, over, say, the next 70 years. Then, by taking 76 as the middle figure, the computer program was run for $n=76$ to estimate a different matrix.

Comparing this matrix with the earlier one revealed that the results are not sensitive to such a change. For example, for $n=73$ the discount factor at a 10 per cent discount rate is 0.137 for the years between 56 and 73; for $n=76$ it is 0.132 between the years 54 and 71, and 0.131 between the years 72 and 76. This change does not make any difference to the shape of the modified discounting method discount factor curve in Figure 4.1. However, the computer program allows analysts to use any figure for n up to 200, and in this way different matrices can be estimated.

A CRITICISM OF THE MODIFIED DISCOUNTING METHOD

At this point it should be made clear that the modified discounting method, which challenges all the conventional cost–benefit analysis criteria, has some critics. The author believes that this is a healthy development as it will stimulate discussion which will, no doubt, bring greater insights into the new method. At this point in time Colin Price (1984b, 1987) appears to be the most persistent critic. It is quite likely that the discussion on modified discounting will continue for some years. The following arguments are extracted, with some modifications, from Price (1984b).

There is no doubt that the modified discounting method represents an improvement on accepted procedures of the net present value calculations. It does so by giving equal weight to all future generations and by circumventing the consequences of normal discounting by which an indefinite stream of future benefit, or cost, is mindlessly reduced to a minute present value. However, some problems have accompanied the attempt to produce a legitimate compromise between the interests of future generations and the apparent preference of present individuals for consumption earlier rather than later in their lifetime.

Let us start with an example from forestry appraisal. Take a site capable of yielding saw-quality oak over a single long rotation. For the sake of simplicity it is assumed that the trees are widely spaced and will not undergo thinning. Maintenance costs are also excluded. The cash flows of several alternative rotations, all with an establishment cost of £1000, are described in Table 4.3.

The initial step is to determine the optimum rotation — that giving the maximum value for the modified discounting method — which will also determine whether the forestry investment should be undertaken. Because the discount factor does not diminish beyond 73, eventually the forest yields sufficient revenue, even after discounting, to offset the establishment cost. The optimal rotation is 200 years, presumably the time at which net physical decay sets into the crop. Note that, using the conventional net present value method, profitability is never achieved, and thus the project is never undertaken.

So far the modified discounting method has given the expected and desired results. However, look at the decision on when to terminate the rotation from the perspective of the forest manager eventually charged with either felling the crop or postponing the decision for, say, 10 years. It will be beneficial to

Table 4.3: Cash flows of alternative rotations

Rotation (years)	Revenue	Conventional net present value factors	Results by means of the net present value method (£)	Modified discount factors	Results by means of the modified discounting method (£)
50	500	0.008 5	−996	0.138 5	−931
60	1 500	0.003 3	−995	0.137 1	−794
70	3 000	0.001 3	−996	0.136 9	−589
80	5 000	0.000 5	−998	0.136 9	−316
90	7 500	0.000 2	−999	0.136 9	27
100	10 000	0.000 1	−999	0.136 9	369
120	14 000	0.000 0	−1 000	0.136 9	917
180	20 000	0.000 0	−1 000	0.136 9	1 738
200	20 500	0.000 0	−1 000	0.136 9	1 806
220	20 000	0.000 0	−1 000	0.136 9	1 738

Source: Price (1984b).

fell as soon as the revenue from felling 10 years hence multiplied by the discount factor for 10 years, 0.41690, becomes less than the revenue from felling now, a point reached at 60 years. But if this is the predicted felling, assuming that future decision-makers will adopt the modified discounting method, then the result is negative and the original investment decision incorrect.

The inconsistency between decisions on rotation length and whether to plant or not arises from the different relativities between time periods, according to whether they are viewed from present or future. Viewed from present, year 70 is hardly discounted at all relative to year 80, so it is profitable to extend the rotation through these periods. However, by the time year 60 has become the point of reference, year 70 is heavily discounted relative to year 60, and even rapid growth of revenue from the forest is insufficient to offset this. The fact of the matter is that, although all distant future time periods beyond 73 years are treated equally in relation to each other, they are not treated equally in relation to the present or the near future. This seems contradictory in view of Kula's assertion that the modified discounting method will allow the decision-maker to treat all generations, present and future, absolutely equally. The assertion, and indeed the whole method, warrants a closer inspection.

As an initial assumption, let us take it that discounting by individuals of their own future consumption is perfectly rational and will lead to efficient allocation of benefit over their own lifetime. Furthermore, consider that future benefits accrue equally, as Kula assumes, to each age class in the population. Suppose that our oak forest has been planted as a community decision, with an intended rotation of 60 years. A young person who acceded to the planting decision expects his share of revenue to be £1500/73, the discounted value of which is £0.07. Yet for a person born in year 50, who has to wait only 10 years for his share, the value of the share is £7.92. In effect, the first person is giving a very substantially greater weight to consumption by a person yet to be born than to himself. He would, therefore, require greater compensation for the elimination of the other's share than for his own and would judge a plantation superior if his own share of benefits were assigned to the other: remarkable altruism, which none the less would apparently wither as the date of felling approached! In this circumstance the method seems to discriminate against benefits to present generations, whereas the net present value method would treat all benefits at year 60 equally.

One might attempt to adopt the method by enforcing the same range of weightings, 1 to $1/(1.10)^{73}$, for each individual, regardless of the remaining length of lifespan. This modification still provokes one major objection. It violates the individual's own assessment of discount factors. In the extreme case for a person aged 72 years there will be a 105 000 per cent discount rate. However, this objection can be overcome if benefits are discounted to the beginning of the individual's life instead of to the beginning of the project period. But in this case, where benefits are equally distributed the discount factor for each point in time is identical, which is to say that discounting is not undertaken at all. Thus the modified discounting method becomes an unnecessary overelaboration.

The attempt to reduce intergenerational injustice while taking account of individual discounting fails to achieve complete equity. There seems to be a fundamental conflict between intergenerational justice and social discounting. For an individual, discounting can be held rational, in that there is a finite risk of death over a given period. This reasoning does not apply to society as a whole. It is also difficult to justify discounting on the grounds of pure time discounting, i.e. preference for benefit merely because it accrues earlier. If individuals prefer to consume now rather than in the future, they would also prefer to consume now rather than to have consumed in the past. This is consistent with the general preference for consuming now rather than at any other time. It is not consistent with the general preference for consumption earlier rather than later. Since each point in time is equally 'now' only once, there is no reason for consumers to prefer consumption at one point in time simply because it is earlier (Price, 1978, 1981, 1984a).

As for discounting on the basis of the diminishing marginal utility of increasing per capita income, there may be sufficient justification. Kula (1984b) reports a widespread counter-argument that the welfare of future generations will be limited by environmental degradation. This could be an argument for equal weight being given to later income on the grounds that financial benefits are required as a substitute for environmental ones. If, however, material goods do not replace environmental ones, then it is better explicitly to predict the utilities of material goods and environment, rather than lump all together as general income. In the latter case, equity may require us to give more weight to the marginal utility imported by material goods accruing to those for whom the environment is a reduced source of welfare.

In summary, Kula's modified discounting method can be classified as a brave effort to resolve a key issue in public investment appraisal. Attempting to accommodate conventionally defined pure time preference, however, restricts the extent to which intergenerational equity can be achieved, because of the collective nature of public investments. Only by abandoning pure time preference discounting can complete intergenerational equity be achieved. To the extent that some discounting can legitimately be based on diminishing marginal utility, it should be applied over the whole timespan of the project, using the normal net present value criterion.

A DEFENCE OF THE MODIFIED DISCOUNTING METHOD

Price's suggestion is that the use of the net present value method without a pure time discount rate will neither achieve intergenerational justice nor create efficiency in public sector economics (Kula, 1984c). His numerical example makes it clear that, if the forestry project is undertaken on a rotation basis, then there will be large benefits accruing to society at each rotation, in perpetuity. In this case an endless number of generations will benefit. Unfortunately the project never takes off by means of the net present value criterion, even with an extremely low and pure time preference free 2.5 per cent discount rate. In this way Price's analysis becomes self-defeating.

However, this example, and the argument that follows it, provoke two objections. First, the figures employed are wildly unrealistic. Take a similar example with different and somewhat more realistic figures in which the oak forest matures around year 110, which is consistent with the normal yield tables for many classes as estimated by the Forestry Commission (1971). Let the establishment cost be £1000, for all rotations, as in the case of Price's example; and let the social time preference rate be 10 per cent also, as in the case of Price's example (see Table 4.4).

Using the conventional net present value criterion, profitability is never achieved. The project qualifies under the rules of the

Table 4.4: Cash flows of alternative rotations

Rotation (years)	Revenue	Conventional net present value factors	Results by means of the net present value method (£)	Modified discount factors	Results by means of the modified discounting method (£)
50	500	0.008 5	−996	0.138 5	−931
60	1 200	0.003 3	−996	0.137 1	−835
70	3 100	0.001 3	−996	0.136 9	−576
80	8 000	0.000 5	−996	0.136 9	95
90	20 000	0.000 2	−996	0.136 9	1 738
100	50 000	0.000 1	−995	0.136 9	5 845
110	120 000	0.000 0	−1 000	0.136 9	15 428
120	119 000	0.000 0	−1 000	0.136 9	15 291

Source: Kula (1984c).

modified discounting method, and the optimal rotation from the point of view of both the community and the forester is 110 years, when the net benefits to society become maximum and the problem Price mentioned does not arise. Of course, by using some other arbitrary numbers it is possible to create various scenarios and indeed argue for or against any case.

Secondly, and most importantly, the assertion of 'Look at the decision on when to terminate the rotation from the perspective of the forester eventually charged with either felling the crop or postponing the decision for (say) ten years' fails to grasp the nature and philosophy of the modified discounting method. Unlike the net present value criterion, the ground rules for this method are laid down in the original position by the representatives from all generations, not by the present foresters who are on the earth plane now and at an advantage in relation to future generations. Once a community decision is made under the veil of ignorance, which prevents representatives from knowing their position on the earth plane, then the foresters, like everybody else, should obey the rules and keep the agreed rotation going. The original position is not thought to be a general assembly which includes everyone at one point in time. This would be stretching fantasy too far and the conception would cease to be a natural guide to intuition. It may be helpful to observe that one or more persons could pretend, at any time, to debate this hypothetical situation simply by reasoning in accordance with the appropriate restrictions.

However, there seems to be a valid point that is latent in Price's analysis regarding the early generations who make a sacrifice by undertaking an investment project on a rotation basis which will materialise beyond their lifetime and who in this way start the ball rolling. It will be helpful to follow this argument through by means of a different forestry example.

Instead of an oak forest, take a coniferous forest that generates large enough benefits to qualify under the modified discounting method and that was planted as a community decision 60 years ago on a 60-year rotation and is now ready for harvesting. In fact this is more or less the case in the United Kingdom, as large-scale plantation of trees, mainly coniferous, took place in the 1920s after the establishment of the Forestry Commission in 1918. If the current generations, having reaped the financial benefits resulting from the felling, do not plant a similar forest for the benefit of the future generations, then they will be in the position of exploiting their offspring as well as their forefathers. To put it in a nutshell, it is a case of wanting something for nothing.

Eventually, this forestry example, which highlights the problem of intergenerational distribution of resources, will take us to the very first generation, which makes a sacrifice by starting off the process of capital accumulation. Indeed, this problem is recognised by Rawls (1972), as he argues that the first generation, having no predecessors, would not receive anything in return for their sacrifice and would be worse off. On the other hand, the last generation would not be required to put any capital aside to pass on.

So, if there is a problem with the Rawlsian doctrine, it is not that the present generation is hard done by through the principles of the doctrine, as Price tends to suggest, but rather the earliest generation, which in a spirit of sacrifice starts up a process that benefits everybody but themselves, a case from which we can only learn a lesson of altruism.

Furthermore, Price's arguments, which tend to imply that the modified discounting method does not yield results as efficient as those based on the net present value criterion, are unfounded. The following points may help to clarify the situation. First, if the modified discounting method is used in public sector investment evaluations throughout, then the project that yields the highest consumption benefits will be ranked first. Although British forestry has a much better chance of qualifying as a viable venture by means of the modified discounting method than the net present value criterion, this does not mean that it should replace more beneficial projects in the public sector investment portfolio. Compared with forestry, if some other public sector investment project generates greater consumption benefits to society, then it should be given priority. Secondly, in Price's analysis the diminishing marginal utility of consumption appears to be the only acceptable rationale for discounting future benefits accruing to individuals, and he also claims that only this can combine justice with efficiency. In effect, the modified discounting method already captures the diminishing marginal utility of consumption argument, as the social time preference rate is the appropriate deflator. Almost all models that are developed to estimate the social time preference rate in economics give recognition to this.

Moreover, an interest rate applied over the whole timespan of the project, using the normal net present value criterion in cost–benefit analysis, would rely on one of the following assumptions:

(a) Society resembles a single individual who has an eternal life. This is an amazing assumption which in effect detaches society from individuals who in fact actually make up a society. Can there be a society without individuals? Can there be a forest without trees?
(b) Society consists of individuals, but those who are alive now (they are also the decision-makers) are more important than those who will join the ranks of society at some later date. An interest rate which contains a parameter based on the diminishing marginal utility of consumption would also imply that those individuals who will be born at some future date, say 200 years from now, are going to experience diminishing marginal utility on their consumption starting from now. Take an individual who will be born at the end of the 22nd century and who is one of the many beneficiaries of a public sector investment project which is starting now. Then discounting these future benefits in the conventional fashion indicates that his experience of diminishing marginal utility dates back to the present point in time.

Given the nature of public sector investment projects, the use of the net present value criterion in their evaluation with either low or high rates of discount must surely be an error, and a disappointment to those who have employed it in theory and practice without question (though undoubtedly in good faith).

APPLICATION OF THE MODIFIED DISCOUNTING METHOD TO BRITISH FORESTRY

By now it must be obvious to the reader that the use of the new method in evaluation of public sector projects will yield very different results from those obtained by means of the conventional criteria. The difference will be most marked in projects such as forestry in which the gestation periods are unusually long. Now let us use the modified discounting method in economic analysis of the forestry project mentioned in Chapter 3. The previous five assumptions regarding future timber prices are retained together with 10, 5 and 3 per cent discount rates. The relevant modified discount factors are shown in Appendix II. The results are as follows:

	MDM with 10% interest rate	*MDM with 5% interest rate*	*MDM with 3% interest rate*
Assumption 1	−1047	−470	315
Assumption 2	−1529	−1443	−1167
Assumption 3	−299	1049	2629
Assumption 4	232	2123	4273
Assumption 5	405	2641	4804

Comparing these results with the ones shown on page 69 shows the difference which modified discounting would make in appraisal of public sector forestry. Here the forestry project yields positive results even at a 10 per cent interest rate if the timber prices increase by 1.5 per cent or more over time. At a 5 per cent interest rate the project fails only if the price of timber remains constant or declines.

To conclude, the ground rules that were laid down to evaluate public sector investment projects in the United Kingdom failed to justify forestry on economic grounds and eventually it was made a special case. The net present value and the discounted cash flows criteria are basically private sector investment evaluation methods which were transferred to cost–

benefit analysis in its heyday in the 1960s. Unfortunately, until quite recently nobody questioned their validity and they were taken for granted in almost all theoretical arguments as well as in practical proposals in cost–benefit analysis. Foresters, following the path paved by economists, always thought that the only way to justify forestry projects would be to advocate a very low discount rate for it (Price, 1973, 1976; Helliwell, 1974, 1975).

The use of the modified discounting method in cost–benefit analysis will have a dramatic impact on the public sector investment portfolio. A portfolio constructed by means of the modified discounting method will contain more long-term projects than the ones selected using the traditional criteria.

5

Private Sector Forestry

The main argument presented in Chapter 4 should not be confused with the evaluation of private sector forestry. The modified discounting method, MDM, is a public sector decision-making criterion in which the public agency treats all generations in an equitable manner with regard to discounting future costs and benefits. It should be remembered that in the case of a public sector investment there is no market for the shareholders, who are the general public, to exchange their share of returns resulting from it. In other words, the beneficiaries, who are, in a way, captive individuals, cannot exchange their share of future net returns but have to wait until they have materialised and are distributed. The situation is very different for a private investor, who can undertake even a very long-term investment (such as an oak afforestation project which matures over a 100-year period) and who is allowed through the existence of a market to exchange the returns that will occur after his death into consumption benefits before his death. Therefore, the discounted cash flows, DCF, or the net present value, NPV, criteria are quite suitable in the private sector.

Another popular private sector investment appraisal method is the internal rate of return, which is a twin sister of the discounted cash flows method. The internal rate of return method yields a zero discounted cash flows, or net present value, figure. The formula is:

$$\frac{B_0 - C_0}{(1+x)^0} + \frac{B_1 - C_1}{(1+x)^1} + \frac{B_2 - C_2}{(1+x)^2} + \ldots + \frac{B_n - C_n}{(1+x)^n} = 0$$

$$\sum_{t=0}^{n} \frac{1}{(1+x)^t} (B_t - C_t) = 0$$

where

t = time, normally expressed in terms of years, 0, 1, 2, ... n
n = project's life
B_t = benefit at time t
C_t = cost at time t
x = internal rate of return (unknown) which satisfies the equation.

Having calculatd the internal rate, the investor compares it with some predetermined interest rate, usually the market rate. If the internal rate proves to be greater than the predetermined rate, then the investment project becomes acceptable. The acceptability condition is:

$$x > r$$

where r = market interest rate.

If the internal rate of a project were, say, 10 per cent, and if the market interest rate were 15 per cent, then there is no point in putting money into this project as more could be obtained by means of a bank deposit. Conversely, if the internal rate of a project were 15 per cent and the market rate were 10 per cent, then the project under consideration makes a great deal of sense.

Note that the internal rate of return formula is a polynomial equation of n degree, which has n roots. That is, there are n rates to satisfy the equation. Therefore, in calculating the internal rate of return we may run into two problems. First, finding the roots of a high degree polynomial can become difficult. Secondly, even if the roots are given, it may be quite difficult to identify the true figure. After all, we want only one real root. If all the roots to a 50-degree polynomial are known and among these a few appear to be plausible, deciding which one to pick could be a problem. (There is a long-winded argument in the economic literature about identifying the real root. See Soper

(1959), Wright (1963), Feldstein and Fleming (1964), and Dasgupta and Pearce (1972).)

In practice, most analysts use the iterative method, which is also called the trial and error method. In this, we start off with an arbitrary rate, say 5 per cent, to see if this yields a zero net present value figure. If not, we try another rate, and another, until we get the correct result.

Example 5.1:

Costs and benefits of an investment project are given below. Find the internal rate of return by using the iterative method.

	Year 0	*Year 1*	*Year 2*	*Year 3*
Estimated costs	300	30	39	47
Estimated benefits	0	140	160	180

To begin with let us choose a 5 per cent figure to see whether this yields a zero net present value. That is:

$$-\frac{1}{(1+0.05)^0}(300)+\frac{1}{(1+0.05)^1}(110)+\frac{1}{(1+0.05)^2}(121)$$

$$+\frac{1}{(1+0.05)^3}(133),$$

$$-(1)(300)+(0.952)(110)+(0.907)(121)+(0.864)(133),$$

$$-300+104.7+109.7+114.9,$$

$$29.3 \neq 0.$$

So, 5 per cent did not yield the desired result. Next we try a higher rate, say 10 per cent. That is:

$$-\frac{1}{(1+0.1)^0}(300)+\frac{1}{(1+0.1)^1}(110)+\frac{1}{(1+0.1)^2}(121)$$

$$+\frac{1}{(1+0.1)^3}(133),$$

$$-(1)(300)+(0.909)(110)+(0.826)(121)+(0.751)(133),$$

$$-300+100+100+100=0$$

So the internal rate of return of this project is 10 per cent.

Using the internal rate of return in preference to the discounted cash flows method has some advantages. First, the internal rate of return gives concise information about the profitability of projects under consideration. For example, if the internal rate of return of a project is, say, 10 per cent, then the investor will be justified in thinking that money invested in this project will grow by 10 per cent, compound, over time. On the other hand, if the person were told that the net present value of this project is £750000 by using, say, an 8 per cent discount rate, he would be left wondering at what rate exactly the money would grow if it were invested. Secondly, the internal rate of return enables the investor to compare the profitability of projects with the market rate of interest.

In view of the above-mentioned advantages the internal rate of return method will be used in this chapter in future discussions.

GRANT AID

Forestry authorities in the British Isles, i.e. the Forestry Commission in Britain, the Forest Service in Northern Ireland, and the Forest and Wildlife Service in the Republic of Ireland, have grant schemes to encourage the uptake of afforestation projects by private investors so that domestic timber output can be increased. These schemes get modified from time to time in response to changing economic circumstances. For example, inflation erodes the real value of the cash grant which reduces the incentive for private foresters. This may necessitate an increase in the nominal value of the grant.

In the UK at present, grants are available in respect of individual areas of 0.25 hectares and over. It is not accepted for individual areas of less than 0.25 hectares to be aggregated. The areas accepted for planting may generally be of any shape provided they harmonise with the landscape. Grants are available at the same rate and under the same terms for both planting and restocking. They will also be paid for the rehabilitation of existing unproductive woodlands. The grants for broadleaved trees are higher than those given for conifers, and encouragement is given to the use of the former in sites which are suitable for such species. In some cases the forestry authorities encourage a mixture of broad-leaved trees and conifers for conservation purposes. Where mixtures are planted, grants will be paid at the appropriate rates in proportion to the area occupied by conifers and broad-leaves respectively at the time of planting.

The rates of grants for Great Britain and Northern Ireland, which are subject to periodic review, are given in Table 5.1. These grants are paid in instalments, the first 80 per cent being payable when the approved area or part thereof has been satisfactorily planted. The balance of 20 per cent will be payable 5 years later, provided that the plantation has been satisfactorily maintained. An inspecton of the land will be carried out prior to planting, and a further inspection will be undertaken as soon as possible after the applicant has notified the forestry authority in charge that planting has been completed. The applicant is expected to maintain the plantation to the forestry authority's satisfaction. In the event of his failing to do so, action may be taken against him in the following ways.

(a) Payment of any instalment in respect of the plantation may be disallowed.
(b) Payment of any instalment in respect of a plantation may be deferred until the applicant has carried out the required maintenance.
(c) The total amount of the instalments already paid in respect of the plantation, including interest, may be recovered.

Applicants are also required to allow officers of the forestry authority to enter on to the land at any reasonable time to consider proposals for work to be done or to inspect areas.

Table 5.1: Forestry grants in the UK as at December 1985

Areas of plantation	Conifers (per hectare)		Broad-leaves (per hectare)	
	Great Britain	Northern Ireland	Great Britain	Northern Ireland
0.25 ha to 0.9 ha	£600	£630	£830	£850
The next 2 ha up to 2.9 ha	£480	£420	£700	£600
The next 7 ha up to 9.9 ha	£400	£315	£600	£500
10 ha and above	£230	£210	£450	£400

Sources: Forestry Commission in Great Britain, Edinburgh, and the Forest Service in Northern Ireland, Belfast.

In Britain, application forms can be obtained from the Forestry Commission's head office in Edinburgh, or from any of the regional conservancy offices in North England (York), East England (Cambridge), West England (Bristol), Wales (Aberystwyth), North Scotland (Inverness), Mid Scotland (Glasgow), and South Scotland (Dumfries). In Northern Ireland they are available at the Forest Service's headquarters in Belfast, or from one of the District Forest Offices in Limavady, Ballymena, Castlewellan, Cookstown, Omagh and Enniskillen.

On receiving an application the forestry authorities first ensure that the land is suitable for forestry. Then they may also undertake consultation with the appropriate agriculture department, local authority or other authorities concerned, on conservation, land use and amenity aspects of the applicant's proposal.

In a social analysis, grants and other subsidies are excluded from calculations simply because they are transfer payments. In other words, by means of grants or subsidies the government transfers income from one sector of society to another. But in private sector analysis, which is the case here, grants form part of the investor's income and thus they are to be taken into consideration. As explained above, the size of grant aid depends on the size of the plantation (see Table 5.1). Therefore, in a private rate of return analysis it is essential to relate calculations to a specific size and location.

A RATE-OF-RETURN ANALYSIS OF PRIVATE FORESTRY

In this section private rate-of-return figures will be calculated for two afforestation projects located in two different grades of soil in Northern Ireland. Details of these sites are shown in Table 5.2. The purpose of choosing two very different locations is to show the reader how different the returns can be in forestry.

Site 1 is idle land and therefore there is no opportunity cost in land use. In other words, the rental value of the land is zero, or nearly zero. This type of land, which is not suitable for traditional agriculture, is common in parts of Northern Ireland, especially in the exposed west. Here the farmer who owns the land decides to plant trees, for which he receives technical advice and cash aid from the Forest Service. Site 2 is a fertile piece of land which is to be taken out of traditional agriculture.

Table 5.2: Details of private afforestation projects in Northern Ireland

Site type	Site details	Site type reference	Assumed block size (ha)	Species	Yield class	Length of a single rotation	Alternative land use
Upland blanket peat, own property	Deep, soft, waterlogged peat	1	100	Sitka spruce	10	50	None
Scrub-covered site, leased property	Light scrub (under 4 m length)	2	100	Sitka spruce	24	35	Beef breeding cows

Cost and output details for both sites are realistic as they are obtained by way of field research.

Site 1: Costs

As indicated by Table 5.3, no cost is incurred by way of scrub cutting as this type of land does not contain scrub vegetation. Drains are cut by tractor-driven ploughs and 310 metres of cut is assumed for each hectare, a total of 35 000 metres for the whole area. Machinery cost amounts to £3300, which is based on the Forest Service's cost data. Total labour cost involved in the draining operation is £900 and this is based on a £0.048 figure per standard minute rate. The total drainage cost is £3200. Ploughing cost is £4400 in total, £3000 of which is machinery cost and £1400 labour. As regards the planting cost, the total sum is calculated to be £18 300 (£3600 labour, £700 machinery and £14 000 materials). In calculating the fencing cost the site is assumed to require 4000 metres of fencing material. Labour and material requirements are £900 and £2600, respectively, a total of £3500 for the whole area. No weeding is required in this area because the soil type on this site does not support weeds. All these expenditures are incurred in year 0.

Another operational assumption is that there will be no brashing. Nowadays brashing is not practised in new plantations in Northern Ireland. Cleaning takes place on inspection racks in year 20. The cost, labour only, amounts to £400 altogether. The

Table 5.3: Cost details of plantation, site 1

Operation	Year	Cost (£) 1985/86 prices each round
Ground preparation:		
Drains	0	44 000
Ploughing	0	4 400
Planting	0	18 300
Fertiliser	0	11 800
Fencing	0	3 500
Maintenance:		
Cleaning	20	400
Fertilising	8, 12, 16, 20, 24, 30	5 700
Drains	15, 25, 40	1 700

use of fertiliser is necessary for this site. The first aerial fertilisation takes place in year 1, immediately after planting and then at regular intervals every 4 years from year 8 until year 30. The cost of the first year's fertiliser is £11800 which consists of £4800 labout and £7000 material (75000 kg fertiliser including delivery). The cost of regular fertiliser is £5700 each round for the whole site. This takes up 25000 kg fertiliser every time it is carried out. Maintenance for drains takes place in years 15, 25 and 40, and costs £1700 each time. No maintenance is assumed for the fences. As regards road construction, since this is a small individual plantation there is no need to construct new roads and the existing ones are assumed to be sufficient for access. As mentioned earlier, the rent for this site is assumed to be zero as there is no alternative use for the site. Table 5.3 shows all the cost details of this project in terms of 1985/86 prices.

Site 1: Benefits

There are three types of cash benefit: grant-in-aid given in two instalments by the Forest Service, yield from thinning, and yield from felling at the end of rotation, year 50.

Grants. The planting grants for this 100-hectare site are calculated as follows:

First 0.9 hectare	=	£567
Next 2 hectares	=	£840
Next 7 hectares	=	£2205
Remaining 90 hectares	=	£18921*

This gives a total of £22533. The first 80 per cent instalment, which is given on plantation (year 0), is £18026, and the second instalment, 20 per cent, given in year 5, is £4507.

Yield from thinning. Thinning starts in year 25 and continues at 5-yearly intervals until the felling, i.e. 25, 30, 35, 40, 45. The output details resulting from thinnings are given in Table 5.4.

Yield from felling. Table 5.4 also shows the output that results from clear felling at the end of year 50.

*£212 × 0.1 + £210 × 90 = £18921.

Table 5.4: Output details for site 1

Yield class (YC), Sitka spruce	Output, m³, thinning, YC10 Years					Felling
	25	30	35	40	45	50
Pulpwood (cmtd 7)	3 000	3 400	3 300	3 200	3 000	15 300
Palletwood (cmtd 18)	100	100	200	300	500	12 200
Saw-wood (cmtd over 24)	0	0	0	0	0	3 300
Total	3 100	3 500	3 500	3 500	3 500	30 800

Source: Forestry Commission (1971).

In order to convert the output into money benefits we need to know the price of timber. It is assumed that the entrepreneur sells his timber standing. The prices for standing timber per m^3 in 1985/86 were:

Pulpwood, cmtd around 7	=	£5.10
Mainly palletwood, cmtd around 18	=	£10.00
Saw-wood, cmtd over 24	=	£14.10

We have all the ingredients for a financial study. Table 5.5 shows the cash flow of this investment in terms of constant

Table 5.5: Cash flow for site 1, 1985/86 prices, Northern Ireland

Years	Cost (£)	Benefits: Grant aid (£)	Benefits: Thinnings (£)	Benefits: Felling (£)	Net benefit (£)
0	82 000	18 026	—	—	− 63 974
5	—	4 507	—	—	+ 4 507
8	5 700	—	—	—	− 5 700
12	5 700	—	—	—	− 5 700
15	1 700	—	—	—	− 1 700
16	5 700	—	—	—	− 5 700
20	6 100	—	—	—	− 6 100
24	5 700	—	—	—	− 5 700
25	1 700	—	16 300	—	+ 14 600
30	5 700	—	18 340	—	+ 12 640
35	—	—	18 830	—	+ 18 330
40	1 700	—	19 320	—	+ 17 620
45	—	—	20 300	—	+ 20 300
50	—	—	—	246 560	+ 246 560

1985/86 prices. The major cost is incurred in year 0 when the site is planted. About 22 per cent of it is covered by the government grant. The second instalment of the grant is in year 5, £4507, which is about 5.5 per cent of the establishment cost. In other words, in this project about 27.5 per cent of the initial cost is covered by the government.

Apart from the large initial cost there are periodic maintenance costs as shown in column 2 of Tables 5.3 and 5.5. The first set of benefits arise in year 25 when thinning starts. Large benefits, £246 560, arise at the end of the rotation, year 50. Obviously there is a very long time interval between costs and benefits in this venture. The net benefit profile of this project is shown in Figure 5.1.

Now we are in a position to find the internal rate of return of this project. For this we use the formula explained at the

Figure 5.1: Net cash benefits of investment, site 1

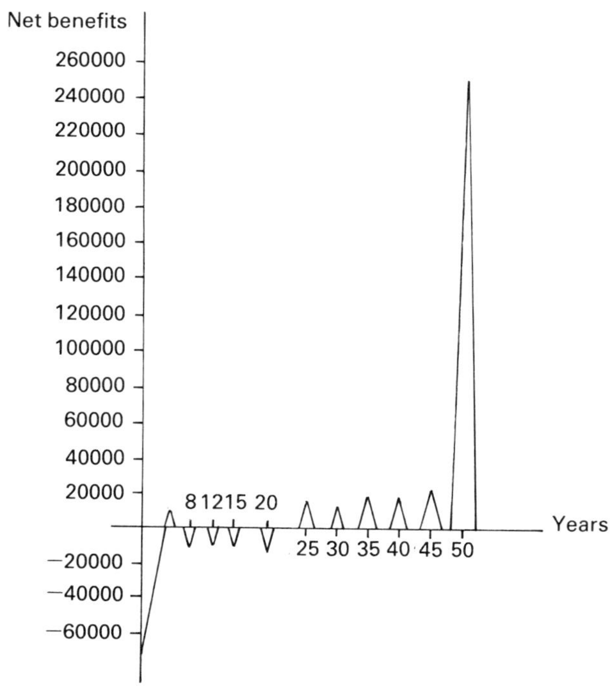

beginning of this chapter. Once again the internal rate is the one that yields zero net present value. That is,

$$\frac{-82000+18026}{(1+x)^0}+\frac{0}{(1+x)^1}+\frac{0}{(1+x)^2}+\cdots+\frac{4507}{(1+x)^5}$$

$$+\cdots+\frac{-5700}{(1+x)^8}+\cdots+\frac{-5700}{(1+x)^{12}}+\cdots+\frac{-1700}{(1+x)^{15}}+\frac{-5700}{(1+x)^{16}}$$

$$+\cdots+\frac{-6100}{(1+x)^{20}}+\cdots+\frac{-5700}{(1+x)^{24}}+\frac{-1700+16300}{(1+x)^{25}}$$

$$+\cdots+\frac{-5700+18340}{(1+x)^{30}}+\cdots+\frac{18830}{(1+x)^{35}}$$

$$+\cdots+\frac{-1700+19320}{(1+x)^{40}}+\cdots+\frac{20300}{(1+x)^{45}}$$

$$+\cdots+\frac{246560}{(1+x)^{50}} = 0$$

Which is the rate that satisfies this equation? In order to find the answer let us use the iterative method which is explained below. In this respect let us start by using a 5 per cent interest rate to find out whether this yields a zero net present value.

$$\frac{-63974}{(1+0.05)^0}+\frac{0}{(1+0.05)^1}+\frac{0}{(1+0.05)^2}$$

$$+\cdots+\frac{4507}{(1+0.05)^5}+\cdots+\frac{-5700}{(1+0.05)^8}+\cdots+\frac{-5700}{(1+0.05)^{12}}$$

$$+\cdots+\frac{-1700}{(1+0.05)^{15}}+\frac{-5700}{(1+0.05)^{16}}+\cdots+\frac{-6100}{(1+0.05)^{20}}$$

$$+\cdots+\frac{-5700}{(1+0.05)^{24}}+\frac{14600}{(1+0.05)^{25}}+\cdots+\frac{12640}{(1+0.05)^{30}}+\cdots.$$

$$+\cdots+\frac{18\,830}{(1+0.05)^{35}}+\cdots+\frac{17\,620}{(1+0.05)^{40}}+\cdots+\frac{20\,300}{(1+0.05)^{45}}$$

$$+\cdots+\frac{246\,560}{(1+0.05)^{50}}$$

That is:

$$\begin{aligned}&(1)(-63\,974)+\cdots+(0.78)(4507)+\cdots+(0.68)(-5700)\\&+\cdots+(0.56)(-5700)+\cdots+(0.48)(-1700)\\&+(0.46)(-5700)+\cdots+(0.38)(-6100)\\&+\cdots+(0.31)(-5700)+\cdots+(0.30)(14\,600)\\&+\cdots+(0.23)(12\,640)+\cdots+(0.18)(18\,830)\\&+\cdots+(0.14)(17\,620)+\cdots+(0.11)(20\,300)\\&+\cdots+(0.09)(246\,560),\end{aligned}$$

which is:

$$\begin{aligned}&-63\,974+3515-3876-3192-816-2622-2318\\&-1767+4380+2907+3389+2467+2233+22\,190,\end{aligned}$$

which is:

$$-78\,565+41\,081\neq 0$$

So, 5 per cent failed to yield the desired result, i.e. zero net present value. We need a lower interest rate for the next iteration. Let us try 3 per cent.

$$\begin{aligned}&(1)(-63\,974)+(0.86)(4507)+(0.79)(-5700)\\&+(0.70)(-5700)+(0.64)(-1700)+(0.62)(-5700)\\&+(0.55)(-6100)+(0.49)(-5700)+(0.48)(14\,600)\\&+(0.41)(12\,640)+(0.36)(18\,830)+(0.31)(17\,620)\\&+(0.26)(20\,300)+(0.23)(246\,560),\end{aligned}$$

which is:

$$-63974 + 3876 - 4503 - 3990 - 1088 - 3534 - 3355$$
$$- 2793 + 7008 + 5182 + 6779 + 5462 + 5278 + 56709$$

That is:

$$-83237 + 90294 \neq 0$$

Three per cent does not yield a zero figure but we are very close to it. In effect, the internal rate of return of this project is 3.2 per cent.

Site 2: Costs

The cost–benefit profile of a plantation located in site 2 is quite different from the one located in site 1. On this, ground preparation consists of scrub clearing and draining; no ploughing is required. Labour cost for scrub clearing is £218 per hectare. This involves cutting the scrub which is up to 4 metres tall and spraying the area with chemicals. The material cost is £24 per hectare. The total cost for ground preparation is £24 200 for the whole site. The total drainage cost is £2900.

Planting takes place in year 1, at a total cost of £21 500 (£6800 labour, £700 machinery and £14 000 materials). Fencing requirement is 3000 metres for the entire site. Cost per metre which includes labour and materials is £0.92, which brings the total fencing cost up to £2760. In this plantation there is no need to use fertiliser at all. In order to prevent young trees from being suffocated, weeding is required in years 1, 2 and 3 with chemical methods. Later on, in years 6 and 10, weeding by cutting shoots is essential. The cost of early weeding is £3500 each time for the whole site. It is £16 400 for each of years 6 and 10. Other maintenance costs comprise upkeep for drains in year 18 at a cost of £5800 and cleaning in the same year at a cost of £8600.

The opportunity cost of land is calculated on the basis of foregone net income had the land been used in the next best alternative sector which is beef cows. According to a survey carried out on 35 farms by the Department of Agriculture, in

Table 5.6: Calculation of opportunity cost of land (site 2), i.e. rental value

		Minus fixed costs, 100 ha (£)					
Type of activity	Gross margin, 100 ha	Depreciation of machinery and equipment	Upkeep and running cost of machinery	Farm fuel	Labour, farmer and wife	Miscellaneous	Net income (£)
Beef cows	28 000	6 500	5 500	1 000	6 000	5 000	4 000

Source: Farm Management Standards (1986).

1985/86 the gross margin (revenue minus variable cost) turned out to be £280 per hectare on farms which are classed as good. The fixed costs per hectare are: depreciation of machinery and equipment, £55; farm fuel, £10; building repairs and miscellaneous costs including rates, about £50; labour by farmer and his wife, about £60. Then the foregone net income would be £40 per hectare. For the whole area it is £4000 per annum. This should also be the rental value of the land. Table 5.6 gives the details for the land's opportunity cost, and Table 5.7 shows the entire cost profile for this operation.

Table 5.7: Cost details of plantation, site 2

Operation	Years	Cost (£) 1985/86 prices
Ground preparation:	0	
Scrub cutting	0	24 200
Draining	0	2 900
Planting	1	21 500
Fencing	1	2 760
Weeding	1, 2, 3	3 500
Maintenance:		
Weeding	6, 10	16 400
Draining	18	5 800
Cleaning	18	8 600
Rent	0–35	4 000

Site 2: Benefits

The level of grant-in-aid for this site is the same as the earlier case, a total of £22533. The only difference is the timing of instalments. Here the planting is carried out in year 1, as opposed to year 0 on site 1. Thus, the first 80 per cent instalment, which is £18026, is given in year 1. The second instalment is given 5 years later in year 6, and is £4507.

The output details are given in Table 5.8. By using constant 1985/86 prices we can calculate the net benefit profile of this investment, which is given in Table 5.9. Notice how large the benefits are at the end of the rotation. On the basis of the net benefit stream given in the last column of the table we can now search for a rate which yields zero net present value. The rate is 6 per cent, which is again obtained by means of the iterative method.

So far internal rates of return are calculated on the basis of a

Table 5.8: Output details, site 2

Yield class Sitka spruce	Yc24, output, m³, thinning Years:				Felling
	15	20	25	30	35
Pulpwood (cmtd 7)	3 800	8 000	6 900	4 200	8 600
Palletwood (cmtd 18)	100	400	1 500	3 400	15 900
Saw-wood (cmtd 24)	0	0	0	800	25 600
Total	3 900	8 400	8 400	8 400	50 100

Source: Forestry Commission (1971).

Table 5.9: Net benefit profile, site 2

Years	Cost (£)	Benefits (£)			Net benefits (£)
		Grant	Thinning	Felling	
0	31 100	—	—	—	–31 100
1	31 760	18 626	—	—	–13 134
2	7 500	—	—	—	–7 500
3	7 500	—	—	—	–7 500
4	4 000	—	—	—	–4 000
5	4 000	—	—	—	–4 000
6	20 400	4 507	—	—	–15 893
7	4 000	—	—	—	–4 000
8	4 000	—	—	—	–4 000
9	4 000	—	—	—	–4 000
10	20 400	—	—	—	–20 400
11	4 000	—	—	—	–4 000
12	4 000	—	—	—	–4 000
13	4 000	—	—	—	–4 000
14	4 000	—	—	—	–4 000
15	4 000	—	20 380	—	16 380
16	4 000	—	—	—	–4 000
17	4 000	—	—	—	–4 000
18	18 400	—	—	—	–18 400
19	4 000	—	—	—	–4 000
20	4 000	—	44 800	—	40 800
21	4 000	—	—	—	–4 000
22	4 000	—	—	—	–4 000
23	4 000	—	—	—	–4 000
24	4 000	—	—	—	–4 000
25	4 000	—	50 190	—	46 190
26	4 000	—	—	—	–4 000
27	4 000	—	—	—	–4 000
28	4 000	—	—	—	–4 000
29	4 000	—	—	—	–4 000
30	4 000	—	66 700	—	62 700
31	4 000	—	—	—	–4 000
32	4 000	—	—	—	–4 000
33	4 000	—	—	—	–4 000
34	4 000	—	—	—	–4 000
35	4 000	—	—	563 820	559 820

single assumption that the prices will remain constant. Table 5.10 shows the future prices of standing timber produced in Northern Ireland under various assumptions. Internal rates of return for both projects, site 1 and site 2, are calculated under all these assumptions and results are shown in Table 5.11. In all these calculations the iterative method is used.

Is forestry a profitable investment? In order to find an answer one must compare the forestry rates of return at least with the market rate of interest. Table 5.12 gives details of the bank base rate of interest between 1968 and 1985 in nominal as well as real terms in the UK. It is obvious that the real rate of interest fluctuates sharply and it can even become negative. For example, in 1975 the real rate of interest was minus 15 per cent. The average rate between 1968 and 1985 is minus 0.1, one-tenth of 1 per cent. Compared with a typical bank deposit rate, in the long run the superiority of forestry investments is considerable. Even with the most unlikely assumption that the real price of timber will decline by 1 per cent compound over time, an afforestation project located in a poor area such as site 1 in Northern Ireland yields a real rate of return of 2 per cent.

TAXATION AND PRIVATE FORESTRY

In the UK successive governments have placed great reliance on securing a tax system to encourage private afforestation. Over the years British forestry has evolved into two distinct sectors. One is distinguished by upland forestry which is characterised by large-scale commercial plantations, often on bare land and almost exclusively with a few species of coniferous trees. The other is lowland afforestation which is on a smaller scale and its pattern is restricted by higher land values and its effects on the landscape. Most of the mixed and broad-leaved woodlands are planted in the lowlands, some of which are old and regarded as part of the British heritage. Therefore, their management and conservation represent an entirely different problem.

The intention of the forest tax system is to take into account the nature of forestry investment which is a long-term commitment. It is almost certain that without generous tax exemptions and grant-in-aid nobody would take up afforestation projects in the UK. The taxes of special significance to forestry are income tax, corporation tax, capital farm tax and capital transfer tax.

Table 5.10: Future price of standing timber in Northern Ireland under different assumptions

Assumptions regarding the future price of timber in Northern Ireland	Years							
	15	20	25	30	35	40	45	50
	£	£	£	£	£	£	£	£
Assumption 2 (no change)								
Pulpwood	5.1	5.1	5.1	5.1	5.1	5.1	5.1	5.1
Palletwood	10.0	10.0	10.0	10.0	10.0	10.0	10.0	10.0
Saw-wood	14.1	14.1	14.1	14.1	14.1	14.1	14.1	14.1
Assumption 2 (1% decrease)								
Pulpwood	4.4	4.2	4.0	3.8	3.6	3.4	3.2	3.1
Palletwood	8.6	8.2	7.8	7.4	7.0	6.7	6.4	6.1
Saw-wood	12.1	11.5	11.0	10.4	9.9	9.4	9.0	8.5
Assumption 3 (1% increase)								
Pulpwood	5.9	6.2	6.5	6.9	7.2	7.6	8.0	8.4
Palletwood	11.6	12.2	12.8	13.5	14.2	14.9	15.6	16.4
Saw-wood	13.4	17.2	18.1	19.0	20.0	21.0	22.1	23.2
Assumption 4 (1.5% increase)								
Pulpwood	6.4	6.9	7.4	8.0	8.6	9.3	10.0	10.7
Palletwood	12.5	13.5	14.5	15.6	16.8	18.1	19.5	21.1
Saw-wood	17.6	19.0	20.5	22.0	23.7	25.6	27.6	29.7
Assumption 5 (2% increase)								
Pulpwood	6.9	7.6	8.4	9.2	10.2	11.3	12.4	13.7
Palletwood	13.5	14.9	16.4	18.1	20.0	22.1	24.4	26.9
Saw-wood	19.0	21.0	23.1	25.5	28.2	31.1	34.1	38.0

Table 5.11: Internal rates of return, sites 1 and 2, under different assumptions regarding future price of timber

Assumptions regarding the price of timber	Internal rate of return (%)	
	Site 1	Site 2
1 (no change)	3.2	6
2 (1% decline per annum)	2	4.5
3 (1% increase per annum)	3.5	7
4 (1.5% increase per annum)	5	7.3
5 (2% increase per annum)	5.5	8.5

Table 5.12: Nominal and real base rate of interest in the UK, 1968–1985

Year	Nominal interest rate, base rate (%)	Inflation rate on the basis of the Robert Price Index (%)	Real interest rate (column 2 minus column 3) (%)
1968	7.5	2.9	4.6
1969	8.0	3.7	4.3
1970	7.5	8.0	−0.5
1971	5.5	10.8	−5.3
1972	7.5	10.3	−2.8
1973	10.0	8.0	2.0
1974	12.0	16.9	−4.9
1975	10.5	25.5	−15.0
1976	12.0	13.9	−1.9
1977	10.5	12.5	−2.0
1978	10.0	11.5	−1.5
1979	14.0	13.4	0.6
1980	15.0	18.0	−3.0
1981	14.5	12.0	2.5
1982	11.5	8.5	3.0
1983	10.0	4.6	5.4
1984	10.5	5.0	5.5
1985	12.5	5.9	6.6

Source: *Bank of England Quarterly Journal,* London.

Income and corporation tax assessment of commercial woods

Where woodlands are managed on a commercial basis and with a view to the realisation of profits, the occupier is assessable for income tax under Schedule B unless he has elected in respect of those woodlands for assessment under Schedule D. Schedule B assumes that the owner receives an annual rent equal to one-third of the annual value of the land in its unimproved state, i.e.

without the trees growing on it. Currently this assumed rent is of the order of £3 per hectare and the taxable income is thus £1. Obviously the tax charge under Schedule B is minimal and in some cases may not be collected by the Inland Revenue as the administration cost would probably exceed the tax due. No tax is payable on the proceeds of any sale of timber or on receipt of planting grants.

However, the occupier may elect instead to place his forest under Schedule D, when all profits are treated as trading profits and taxed accordingly in the annual balance between income and expenditure. It is most important to note that under Schedule D all the costs of forest establishment and maintenance can be offset against other sources of income. Also, grants are taxable under this schedule. To make this clear, assume that apart from forestry the farmer who plants trees on site 2 has another business. In afforestation the net cost in years 0 and 1 is £31 100 and £13 134, respectively. He can offset these costs against income accruing to him in the other business for these years. In this way an increasingly valuable asset is being created out of income which would otherwise be subject to high rates of personal income tax. This relief against other taxable income is particularly attractive to high-rate taxpayers, and forestry management companies have been successful in encouraging such people to invest in afforestation projects. However, the vast majority of farmers are not high-rate taxpayers, and so they gain less benefit from these reliefs.

Once the occupier has elected for a particular schedule, this is binding so long as he remains as occupier. For example, election to Schedule D must be notified to the Inland Revenue within two years of the expenditure's being incurred. On a change of ownership, the woodland reverts automatically to Schedule B. There is an option available which allows owner, taxed under Schedule B, to transfer any plantation of less than 10 years old into Schedule D as if it were a separate estate. Corporation tax is paid under the same schedules but at a higher rate, starting at 40 per cent.

Capital gains tax

In computing the taxable gain on disposal of a forestry property, growing timber is left out of account. In other words, the increase in the value of the underlying land is subject to tax, not the timber. Since 1982, in determining the capital gains the effect of inflation has been eliminated from calculations. For example, after 20 years of ownership a landlord is selling his property to another person. Let us assume that over this period the market value of land has doubled nominally. If, during that period, the retail price index, which is a measure for inflation, has also doubled, then it means that the real value of the land has remained constant, the capital gain is zero and no tax is due.

Capital transfer tax

Owners are also required to pay capital transfer/inheritance tax. The payment on the trees may be deferred until the timber is sold. Valuation for assessment is then on the sale proceeds rather than on the valuation at death. The tax on land cannot be similarly deferred though it can be paid in instalments, usually interest free, over ten years. This tax is payable on half its full market value. In March 1986, the capital transfer tax on lifetime gifts was abolished altogether. Other exemptions are a tax-free threshold of £67 000 and no tax when the transfer takes place between husband and wife.

Taxation of non-commercial woods

Woods planted for amenity purposes and maintained as part of a heritage may gain exemption from the capital transfer tax. For this exemption there are a number of conditions. These are: the land must be actively preserved in its character; reasonable public access must be secured; and that the woodlands are not subsequently sold. For further details of taxation see Hart (1987).

RESPONSIVENESS OF PRIVATE SECTOR FORESTRY TO CHANGES IN GRANT AID AND OTHER VARIABLES

What are the factors that determine the size of forestry uptake in the private sector? Surely when a person invests in forestry he must do so simply because he believes that it is a good business.

In a minority of cases, however, tree planting may take place for ornamental purposes. Nearly 90 per cent of all afforestation in the United Kingdom takes place in Scotland and the area under coniferous plantations has increased by about 500 000 hectares in the last 10 years. The private sector was responsible for over 80 per cent of the activity in the Scottish afforestation market in 1986. Almost all recent plantations are composed of Sitka spruce and lodgepole pine.

Since the introduction of the 1980 forestry policy statement, the enhanced rates of grant, and simplified procedures introduced under the Forestry Grant Scheme in 1981, new planting in the private sector has risen sharply. Table 5.13 shows the trend for restocking as well as new planting since 1980/81. In a parliamentary statement on 10 December 1980, it was emphasised that, with the projected rise in demand for timber into the next century and with the world's forests likely to come under increasing pressure, the government believed that long-term confidence in forestry in the United Kingdom was fully justified. There should be scope for new planting to continue in the immediate future at broadly the rate of the past 25 years, while preserving an acceptable balance with agriculture and the environment. The statement made it clear that the government wished to see greater participation by the private sector in new planting. In order to achieve this it was emphasised that the administration of grant aid and felling licensing would be made less complex. A single new scheme was introduced on 1 October 1981, the main features of which are planting grants, a simplified plan of operations and a minimum of legal formalities. In introducing the scheme it was also made clear that the

Table 5.13: Area of planting by private woodland owners for which grants are paid

Year	New planting		Restocking		Total	
	Conifer	Broad-leaf	Conifer	Broad-leaf	Conifer	Broad-leaf
1980/81	7 918	482	2 174	825	10 092	1 307
1981/82	12 068	394	2 583	867	14 651	1 261
1982/83	11 911	600	2 207	911	14 118	1 511
1983/84	15 995	678	2 034	1 033	18 029	1 711
1984/85	15 434	595	2 014	1 052	17 448	1 647
1985/86	17 673	772	1 464	989	18 445	2 453

Source: Forestry Commission (1980–86).

government intended to continue the current income tax arrangements for forestry in order to maintain confidence in the private sector.

One major effect of intensifying afforestation in Britain has been the increase in land prices. In the 1960s forestry land in Scotland was fetching about £10 per acre. By 1970 prices had reached £40 per acre, and had soared to an average of £260 by 1980 (McEwan, 1981). In 1985 new record prices were set with up to £350 per acre being paid in parts of Scotland. Figure 5.2 shows the upward trend in land prices since 1958. Tompkins (1986) argues that Scottish hill farmers value such land in an entirely separate market according to the number of sheep it will support. A piece of land which is fetching about £300 per acre in forestry would fetch only about £50–150 per acre for a farmer for sheep grazing.

The response of private foresters to an increase in the grant levels must be positive otherwise there would be no point in

Figure 5.2: Average land prices in forestry, Great Britain, 1958–84

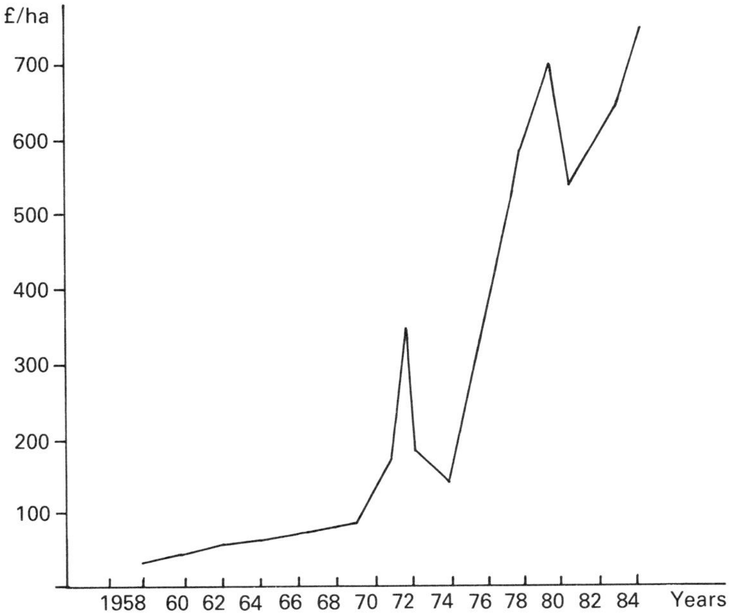

Source: Syson (1985).

increasing grants, or indeed in introducing them in the first place. Other variables which may have a similar impact on the uptake are likely to be tax incentives and the price of timber, which are an indication of profitability of afforestation. All these variables should normally be positively correlated with decisions to plant trees.

On the other hand, there must be other variables which could discourage potential investors in forestry. The real rate of interest must be one of these factors. If this rate is, say, increasing the negative effect will be twofold. First, the high interest rate will make the establishment cost dearer (interest rate is the cost of borrowing), thus discouraging potential investors. Secondly, with high interest rates investments in time deposits will become rather more attractive, and forestry, and other investments in general, will be less attractive. Another negatively correlated variable is likely to be the price of land, the reason being that afforestation reduces the marketability of land. For example, in a period of rising land prices landowners may prefer to sell or lease their land. This may, however, prove difficult if they are tied up in afforestation. Prices of alternative agricultural commodities, which could have been produced had the land not been put under trees, are likely to be another negatively correlated variable. A high price for these commodities would make agriculture more lucrative in comparison with forestry.

Constructing planting functions for private afforestation projects in Northern Ireland

I should like to warn those who are not familiar with the modern econometric technology that this section may prove to be rather difficult to follow. My advice is that they should avoid the mechanics of the calculations involved but concentrate on the final results and their meanings which are explained at the end of this chapter.

Northern Ireland, as part of the United Kingdom, has some advantages and disadvantages in forestry compared with Britain. The most obvious advantage is its mild and moist climate and soil types which are conducive to afforestation. Nowhere in Europe do trees grow faster than in Ireland. It takes 35–50 years to grow Sitka spruce there. The average growth periods for similar species are about 50 years in

Scotland, 80 years in Germany and the Benelux countries and 90 years in most parts of Scandinavia (Kula 1986b, c). Another advantage is the price of land, which is not excessive; this was emphasised in a recent report published in *Forestry and British Timber* (1985). This whetted the appetites of potential investors who attended a number of promotional campaigns in Belfast (*Forestry and British Timber*, 1986).

On the debit side, land in Northern Ireland is fragmented into small ownerships which makes acquisition rather difficult. In order to achieve economies of scale, investors are keen to buy sizeable plots which are difficult to come by in Ulster. Furthermore, Irish farmers are unwilling to part with their land as it is regarded as a status symbol in most areas. Another factor is the lack of forestry tradition. This is widespread throughout the United Kingdom, but applies particularly in Northern Ireland.

Parameters of a planting function

Almost all private plantations in Northern Ireland are coniferous, Sitka spruce being the overwhelmingly dominant species. The following model stems from the author's recent work with McKillop (see Kula and McKillop, 1988, and McKillop and Kula, 1987). The planting function in its most general form consists of the following variables:

$$\Delta Q_t = f(PL_t, PS_t, I_t, G_t, R_t, T_t)$$

Where

ΔQ_t is the change in the size of private softwood afforestation in year *t*
PL_t is the price of land in year *t*
PS_t is the price of imported sawn softwood in year *t* which dominates the price in the British Isles
I_t is a weighted price index of selected agricultural commodities in year *t*
G_t is the present value of government grants available for softwood planting in year *t*
R_t is the market rate of interest, in real terms, in year *t*
T_t is the level of tax incentive.

The relevance of all these independent variables must be obvious. Land prices, PL, is included on the basis that afforestation reduces the marketability of land as has already been explained. The correlation between the dependent variable, ΔQ_t and PL should be negative. That is, an increase in land prices is bound to make afforestation projects less attractive for the reason given above.

The price of imported softwood, PS_t, should be positively correlated with the dependent variable ΔQ_t, as it is an indication of the profitability. If the timber prices are rising, this should encourage investors to take up afforestation projects on the grounds that profits in this line of business are increasing.

The independent variable I_t, price index of select agricultural commodities, is relevant because it is a measure of the profitability in the competing agricultural sector. The correlation between this variable and ΔQ_t should be negative because an improvement in agricultural profits would, *ceteris paribus*, reduce the attractiveness of afforestation projects.

The present value of government grants for forestry, G_t, is a crucial parameter in planting functions, and one should expect it to be positively correlated with the dependent variable, ΔQ_t. This variable must be expressed in terms of net present value because of the fact that grants are given in two instalments (see page 62).

Finally, the market rate of interest, R_t and an index based upon tax levels, T_t are also included. However, neither of these proves to be significant for the reasons which will be explained below.

The planting function is specified in log-linear form, which tends to give the best results. It is also very convenient for elasticity calculations. The specification by ignoring the time lags is as follows:

Log-linear model without lags

$$\ln \Delta Q_t = \ln \beta_0 + \beta_1 \ln PL_t + \beta_2 \ln PS_t + \beta_3 \ln I_t$$

$$+ \beta_4 \ln G_t + \beta_5 \ln R_t + \beta_6 \ln T_t$$

However, this model is incomplete owing to the problems of delayed response. For example, the current value of the dependent variable, ΔQ_t, is a function not only of the current grant, but also of the past, or lagged, values of it. Because of this problem a distributed lag model which was first developed by

Almon (1965) and refined by Cooper (1972) is used. In this coefficients are constrained to lie on a polynomial-shaped lag. The lag structure specified in the model was that of a polynomial degree of three. This contains the lag structure, to have at most two turning points. However, as detailed in the results, page 132, the lag structure for the majority of variables was quadratic in form.

Log-linear model with distributed lag

$$\ln \Delta Q_t = \ln \beta_0 + \sum_{i=0}^{n-1} \beta 1_i \ln PL_{t-i} + \sum_{i=0}^{n-1} \beta 2_i \ln PS_{t-i}$$

$$+ \sum_{i=0}^{n-1} \beta 3_i \ln I_{t-i} + \sum_{i=0}^{n-1} \beta 4_i \ln G_{t-i} + \sum_{i=0}^{n-1} \beta 5_i \ln R_{t-i}$$

$$+ \sum_{i=0}^{n-1} \beta 6_i \ln T_{t-i}$$

Figure 5.3 shows the behaviour of the lag structure with respect to government grant, G_t, over a five-year period.

Figure 5.3: Behaviour of the government grant, G_t, with a distributed lag in log-linear planting function

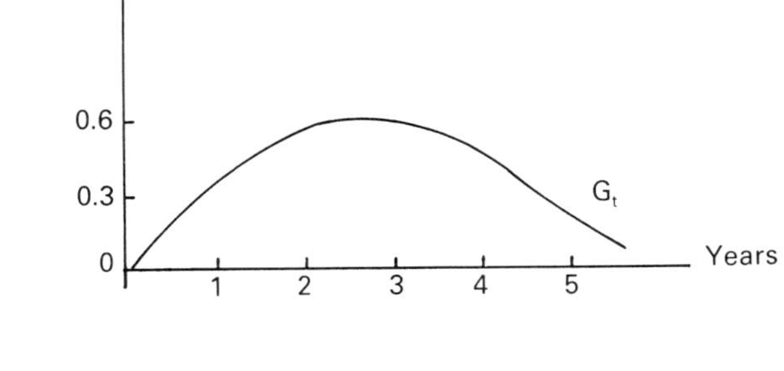

With regard to the relevant lag lengths, i.e. the number of years which the change in independent variables will last, experiments with lag lengths of $n = 4$ to $n = 12$ were conducted. The criteria for deciding upon the optimal lag length are R^2 (the goodness of fit) and a requirement of close similarity of the weights between the optimal lag and lags of slightly longer length. On the basis of these criteria, lag lengths of the order $n = 6$ were arrived at.

The reader should note that, if the length of the lag is incorrectly specified, then the estimates will be biased.

The data sample covers the years 1948/49 to 1983/84, a total of 36 observations with explanatory variables, where appropriate, expressed in 1966/67 prices. Observations for the dependent variable, ΔQ_t, annual increase in softwood plantation, are shown in the second column of Table 5.14. Land prices in Northern Ireland, the first independent variable, are deflated by an index of agricultural product prices in order to capture the change in real terms. The price of imported sawn softwood is deflated by the wholesale price index. A composite index of agricultural commodities, I_t, is deflated by the agricultural price index. This composite index apportioned a 30 per cent weight to cattle and 70 per cent weight to sheep prices on the grounds that the bulk of afforestation takes place on hill and upland areas suitable in the main for sheep farming.

Over the years, the grant schemes have changed many times. For example, the first grants administered by the Forest Service related to ground preparation as well as planting. Furthermore, larger grants were available for corporate bodies than for private individuals. From 1947 to 1981 the main focus for grant aid was the Dedication Scheme, offering a planting grant with an annual management grant attached. Introduced in 1981 and currently in operation is the Forestry Grant Scheme which consists of a planting grant only and no subsequent management grant; 80 per cent is paid on planting and the remainder after five years, subject to satisfactory establishment. To accommodate the time value of money concept it was decided to express the total grant available to an applicant in present value form with future pay-offs being discounted at the average bank base rate for the year under construction.

As regards the average annual interest rate, R_t, on an *a priori* basis one would normally expect an inverse relationship between interest rates and forestry investment. High interest

Table 5.14: Variables for Kula–McKillop planting functions

Year	Increased hectares under softwood afforestation (new planting and restocking) ΔQ_t	Price of land per hectare at constant prices PL_t	Price of imported sawn softwood at constant prices PS_t	Index of agricultural commodities at constant prices I_t	Present value of grant at constant prices G_t
1948/49	81	44.6	79.9	100.5	74.2
1949/50	81	42.8	74.8	100.2	83.7
1950/51	68	47.9	87.4	94.1	88.9
1951/52	62	45.2	101.4	98.4	88.9
1952/53	149	33.9	93.1	105.2	84.1
1953/54	150	30.3	86.3	107.5	85.5
1954/55	162	36.1	87.0	99.4	100.0
1955/56	214	26.0	90.2	94.6	107.9
1956/57	263	19.7	91.2	95.7	102.9
1957/58	197	26.0	84.6	101.3	107.3
1958/59	173	35.7	77.7	99.4	112.7
1959/60	200	42.2	81.9	102.0	110.7
1960/61	220	50.2	88.7	105.3	108.0
1961/62	201	47.1	86.2	107.2	107.2
1962/63	215	64.4	93.7	101.6	106.7
1963/64	214	89.2	99.8	105.8	108.7
1964/65	227	100.4	104.9	107.7	103.5
1965/66	194	98.1	105.7	97.1	102.6
1966/67	187	100.0	100.0	100.0	100.0
1967/68	205	100.8	103.4	106.8	120.2
1968/69	223	100.1	110.0	112.0	137.3
1969/70	283	123.1	113.8	112.8	130.2
1970/71	236	134.3	114.6	118.6	120.3
1971/72	223	147.5	108.6	114.7	109.0
1972/73	157	256.3	115.0	116.5	93.0
1973/74	243	316.9	150.4	129.8	78.0
1974/75	136	153.3	152.1	104.3	63.9
1975/76	73	178.2	129.0	105.9	53.4
1976/77	34	211.9	136.7	109.6	47.4
1977/78	62	234.2	133.1	142.2	79.5
1978/79	24	333.5	123.5	167.3	100.5
1979/80	37	380.8	128.7	150.4	83.0
1980/81	125	248.1	121.5	144.3	71.9
1981/82	181	259.1	105.1	131.1	273.0
1982/83	34	209.6	105.3	114.4	452.0
1983/84	66	236.3	119.7	112.5	421.0

Source: Kula and McKillop (1988).

rates raise both borrowing costs and the opportunity cost of locking finance into forestry. However, this variable did not prove to be significant. Its insignificance may be explained by the fact that initial investment costs are grant aided.

Another variable which did not prove to be significant was that of the tax incentive, T_t. Various attempts to incorporate this variable into regression equations have been unsuccessful. A possible explanation of this is the fact that financial institutions, likely to be more responsive to tax incentives than small farmers, have so far been unwilling to invest in afforestation projects in Northern Ireland. This reluctance to invest is explained when one considers that, in order to achieve economies of scale, financial institutions are interested in large-scale afforestation (100 hectares and over). As mentioned earlier, such tracts of land are not available in Northern Ireland as land is fragmented into small ownerships. The author believes that a similar analysis conducted for Great Britain would result in a highly significant role for tax incentives. The observations for statistically significant variables are shown in Table 5.14.

Results

The regression results with the distributed lag are as follows:

$$\ln \Delta Q_t = 6.3 + \sum_{i=0}^{5} \beta 1_i \ln PL_{t-i} + \sum_{i=0}^{5} \beta 2_i \ln PS_{t-i}$$

$$+ \sum_{i=0}^{5} \beta 3_i \ln I_{t-i} + \sum_{i=0}^{5} \beta 4_i \ln G_{t-i}$$

$$R^2 = 0.9; \quad DW = 2.2; \quad F(12, 23) = 17.6$$

where R^2 refers to the goodness of fit; DW is the Durbin-Watson statistic, which measures the level of autocorrelation, and the F statistic measures the significance of the regression as a whole.

The polynomial lag coefficients, i.e. details of $\beta 1_i$, $\beta 2_i$, $\beta 3_i$ and $\beta 4_i$, are:

$\beta 1_0 = 0.5$ (2.6)	$\beta 2_0 = 1.2$ (2.5)	$\beta 3_0 = -1.6$ (2.6)	$\beta 4_0 = -0.1$ (0.3)
$\beta 1_1 = 0.0$ (0.1)	$\beta 2_1 = 1.1$ (4.3)	$\beta 3_1 = -1.5$ (4.1)	$\beta 4_1 = 0.4$ (3.1)
$\beta 1_2 = -0.2$ (1.7)	$\beta 2_2 = 0.9$ (3.6)	$\beta 3_2 = -1.2$ (3.4)	$\beta 4_2 = 0.5$ (3.6)
$\beta 1_3 = -0.3$ (3.1)	$\beta 2_3 = 0.8$ (3.0)	$\beta 3_3 = -0.8$ (3.3)	$\beta 4_3 = 0.4$ (3.3)
$\beta 1_4 = -0.3$ (2.0)	$\beta 2_4 = 0.5$ (1.8)	$\beta 3_4 = -0.5$ (1.5)	$\beta 4_4 = 0.2$ (1.2)
$\beta 1_5 = -0.1$ (0.9)	$\beta 2_5 = 0.2$ (1.0)	$\beta 3_5 = -0.2$ (0.3)	$\beta 4_5 = 0.0$ (0.2)

The sums of the lag coefficients are:

$$\sum_{i=0}^{5} \beta 1_i = -0.4\ (2.2); \quad \sum_{i=0}^{5} \beta 2_i = 4.7\ (3.7);$$

$$\sum_{i=0}^{5} \beta 3_i = -5.8\ (5.1); \quad \sum_{i=0}^{5} \beta 4_i = 1.4\ (3.7)$$

where the numbers in brackets refer to the *t* statistic, which is a measure of statistical significance.

The meaning of these results is quite straightforward. Take the grant variable, for instance. The elasticity of planting function with respect to this variable is 1.4. This means that a 1 per cent increase in the government grant, in real terms, would generate about a 1.4 per cent increase in private forestry uptake over a 5-year period. The elasticity of the function with respect to variable PS, the price of sawn softwood, is 4.7, meaning that a 1 per cent increase in timber prices would generate an approximate 4.7 per cent increase in plantation. Likewise, if there were a 1 per cent increase in I_t, the price index for selected commodities, this would generate a 5.8 per cent decrease in afforestation.

In terms of 'goodness of fit', the model is a good representation of the underlying data. All coefficients have theoretically expected signs, that is, annual plantation is positively correlated with both timber prices and grant levels and is negatively related to the index of agricultural commodities and land prices. *Vis-a-vis* land prices, as detailed earlier, a negative correlation is to be expected for Northern Ireland. In other regions of the United Kingdom where afforestation is expanding, this negative relationship may not, however, pertain. As explained above, this is primarily because financial institutions, such as pension funds and investment banks which are attracted by grants, tax exemptions and a stable political climate have placed upward pressure on land prices.

The main variable of interest in this study is that of grant aid which can be employed as a policy variable to influence forestry uptake. As can be seen in Figure 5.3, the weighted distribution takes the form of an inverted V distribution: weights increase up to year three, and then decline on moving through the remainder of the time period. That is to say that the maximum impact of an increase in grant will be realised around year three.

It should also be noted that the estimated coefficients for the prices of timber and the index encompassing sheep and cattle indicate that these variables are highly elastic over the period.

This empirical research on Northern Ireland has shown that, according to the Kula–McKillop forestry planting function, only four variables are of relevance, namely, land prices, softwood prices, a weighted price index of agricultural commodities and grant-in-aid. Furthermore, the models employed emphasise the importance of the lag structure in private afforestation projects.

From a policy perspective it is evident that the level of grant can be employed to influence forestry uptake. It must be said, however, that, as uptake is marginally elastic with respect to grant, if the government wishes to encourage substantially the size of private sector forestry, a quite fundamental restructuring of the grant system and/or a revision of the subsidies payable on substitute agricultural commodities must be undertaken.

The response of the private sector to afforestation projects in Northern Ireland has, however, so far been disappointing. It is the author's view that any further restructuring of forestry and agricultural subsidy schemes to reallocate land in the Province should take two added factors into account. First, the absence of a steady income from forestry discourages many potential foresters. It takes an average of about 20 years for the first sizeable benefits to arise from thinnings. Secondly, some Ulster farmers are conservative and may not readily accept the idea of abandoning traditional farming in favour of forestry, irrespective of the level and structure of grants available.

6

The Optimum Rotation Problem in Forestry

One of the most interesting problems in forestry economics is the selection of the optimal time to cut a tree or an even-aged forest. An even-aged forest refers to timber stands composed of trees of only one age group. This problem has been known since the work of Faustmann in the mid-nineteenth century (Faustmann, 1849). As a tree ages, the volume of wood it produces increases up to a certain point, beyond which the tree begins to decay from old age. Figure 6.1 shows the growth cycle of a representative tree. After the tree has become firmly established the volume of wood grows at an increasing rate up to a point α. In terms of mathematics, between points 0 and α, the first and second derivatives of the volume function are positive, i.e. $\mathrm{d}V/\mathrm{d}t$, $\mathrm{d}^2V/\mathrm{d}t^2 > 0$. Beyond α the volume increase continues but at a slower rate until β when the maximum timber output is achieved. That is, between α and β the first derivative of the function is positive, $\mathrm{d}V/\mathrm{d}t > 0$, but the second derivative is negative, $\mathrm{d}^2V/\mathrm{d}t^2 < 0$. At point β, which is the turning point of the volume curve, the derivatives are zero.

In practice the harvesting period depends on many factors such as the final use of the tree, e.g. whether it is for a bean stick, Christmas decoration, pulp, palletwood or saw-wood. Other important factors are the cost of harvesting and the price of wood at the time when cutting is contemplated. For example, the demand for wood in the area may be very depressed at a given time. It is only common sense to postpone the felling until the price is favourable. When the tree is grown for timber, the relevant cycle is usually between α and β. Between 0 and α the tree may be too small to be of value, which may justify ignoring this phase. The period beyond β may also be irrelevant because

Figure 6.1: The growth cycle of a representative tree. Between 0 and α the tree grows at an increasing rate. Between α and β it grows at a decreasing rate. At β the growth is zero and the maximum level of output is achieved

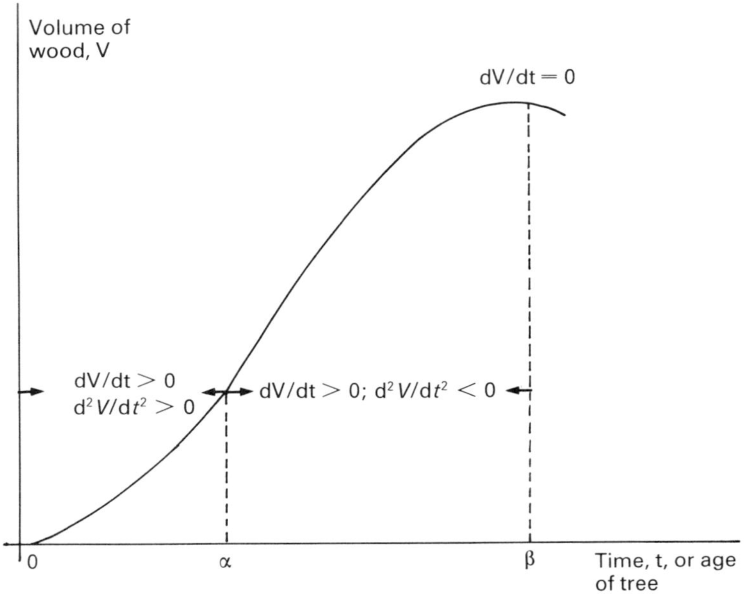

postponing felling beyond this point will only be justified if the anticipated future price increase outweighs the reduction in output and the risks of disease, insect predation and windblow. The growth cycle for a stand of trees of identical age will also have a similar shape to the schedule shown in Figure 6.1, the relevant period being between α and β. Table 6.1 gives the volume increase, per hectare, for three different yield classes of Sitka spruce, YC 12, YC 16 and YC 22, under a no-thinning regime. Figure 6.2 shows the relevant yield curves.

OPTIMUM ROTATION FOR A PRIVATE OWNER

Identifying the optimum rotation requires clear assumptions without which there can be no solution to the problem. In establishing these assumptions we shall also try to simplify the problem without losing the essence of the argument. The assumptions are as follows.

Table 6.1: Growth cycles for Sitka spruce under a no-thinning regime; 1 ha, three different yield classes

Age, t, years	Volume, V, per hectare		
	YC 12	YC 16	YC 22
25	133	207	341
30	214	314	480
35	301	423	615
40	386	526	740
45	465	617	862
50	534	698	967
55	593	770	1065
60	642	834	1145
65	683	891	1220
70	718	940	1271
75	751	982	1326

Source: Forestry Commission (1981).

Figure 6.2: Growth cycle for Sitka spruce stands, per hectare, yield classes (YC) 22, 16 and 12

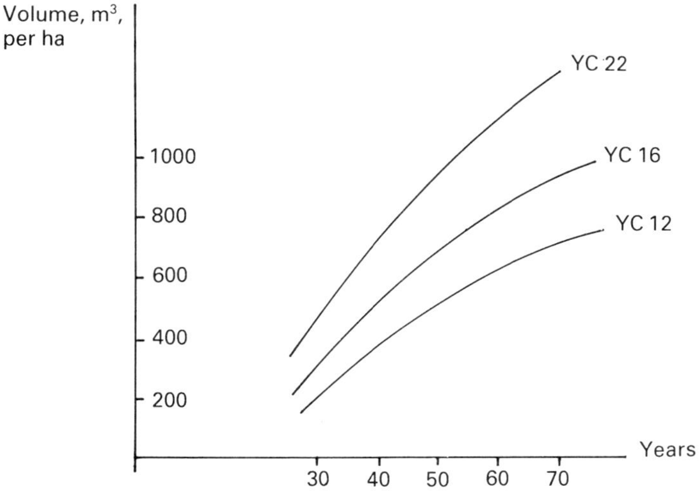

Assumption 1. Future timber yields from a stand that is even-aged are known with certainty, or, to put it another way, the productivity of land on which trees are planted is known. When the output details are worked out, no substantial change is expected in the fertility of the land or in the management technology over time.

Assumption 2. The trees are widely spaced and will not undergo thinning. That is, a no-thinning regime is assumed.

Assumption 3. The trees will be sold standing, i.e. harvesting and haulage costs will not be incurred by the owner.

Assumption 4. All future timber prices are constant and known. The price is assumed not to be sensitive to the age or volume of wood. That is, a cubic metre of wood has the same value, independent of the size and the age of the trees cut down. This is somewhat unrealistic, but at this stage divergence from this assumption will complicate the problem.

Assumption 5. The interest rate is known with certainty and it is constant over all future periods.

Assumption 6. Planting and replanting costs are also known and they are constant for all future periods. When replanting is involved it is done immediately after harvesting the stand. The intervals between planting and replanting are constant.

Assumption 7. Risks from forest fires, windblow and disease are minimal and are thus ignored.

Assumption 8. Forest land can be bought, sold and rented. This price could be very low or very high, depending on the demand and supply conditions for the land.

Assumption 9. There is no taxation of any kind. It is well known to many forestry economists that taxation makes quite a difference to the optimum rotation decision; this will be dealt with later on.

Single-rotation solution

In this the owner tries to identify a time period to maximise his profit over a single cycle. Although the single-rotation model has some serious limitations which will be explained later (p. 141), it is useful to work it out at this stage as it gives an intuitive understanding of the problem.

Bearing the above assumptions in mind, let us argue that the owner has a very simple revenue function:*

$$R_t = A\sqrt{1+t} \qquad \text{or}$$

$$R_t = A(1+t)^{\frac{1}{2}} \qquad (6.1)$$

where $t =$ time expressed in terms of years, 0, 1, 2 ...; $R_t =$ revenue at time t resulting from the sale of timber; $A =$ constant.

Let the initial capital cost be C, and the interest rate or the opportunity cost of capital be r. The net present value of this investment is:

$$\text{NPV} = -C + \frac{1}{(1+r)^t} R_t$$

which can also be written as:

$$\text{NPV} = R_t e^{-rt} - C \qquad (6.2)$$

What is the felling time which maximises the net present value of this project? To find the answer, first we differentiate (6.2) with respect to time t. That is:

$$\frac{\text{dNPV}}{\text{d}t} = -rR_t e^{-rt} + \frac{\text{d}R_t}{\text{d}t} e^{-rt} \qquad (6.3)$$

Then we set (6.3) equal to zero

$$-rR_t e^{-rt} + \frac{\text{d}R_t}{\text{d}t} e^{-rt} = 0$$

or

$$rR_t e^{-rt} = \frac{\text{d}R_t}{\text{d}t} e^{-rt} \qquad (6.4)$$

*It should be noted that a revenue function of this kind is based roughly on the physical growth pattern shown in Figure 6.1. As the stand grows, the size and price pattern of the wood will change, i.e. volume per hectare and price per m^3 will increase with age (see Chapter 3, Figure 3.2). However, it is the author's view that at this stage the use of a more realistic revenue function would complicate the problem.

To solve it for r we divide both sides by $R_t e^{-rt}$, which yields:

$$r = \frac{\mathrm{d}R_t/\mathrm{d}t}{R_t} \tag{6.5}$$

Equation (6.5) tells us that the marginal rate of return on capital must be equal to the interest rate. That is, the net present value of this investment is maximised when the marginal revenue from the selling of timber is equal to the opportunity cost of the capital, i.e. the market rate of interest. In other words, the trees should be felled when the revenue equals the market rate of interest. This should be intuitively obvious. It is not profitable to maintain the stand when the marginal revenue becomes lower than the market rate of interest. It would then be profitable to cut the forest and deposit the timber proceeds in a bank account. In Figure 6.3 the optimum rotation is found geometrically. There are two curves in the figure, the revenue curve which results from the sale of timber, and the opportunity cost of capital (the interest rate curve). Note that the vertical axis is a logarithmic scale so that continuously compounded interest appears as a straight line. Also the revenue curve becomes more compressed. The optimum solution is where the

Figure 6.3: Geometric solution to the optimum rotation problem

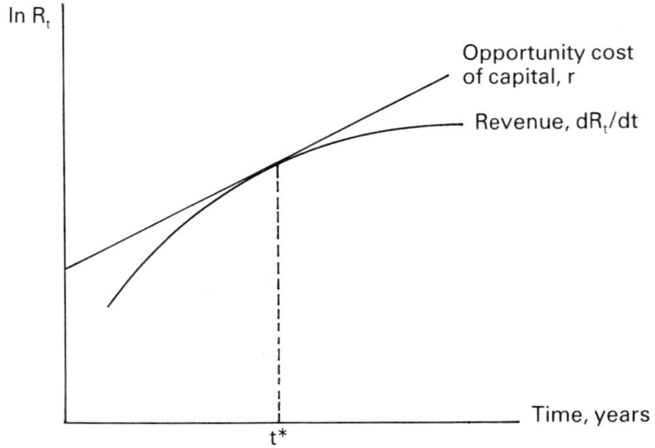

opportunity cost curve, which corresponds to the left-hand side of equation (6.5), becomes tangent to the revenue curve, or it could cut it.

Example 6.1

The revenue function $R_t = 4000\sqrt{1+t} = 4000\,(1+t)^{\frac{1}{2}}$
$r = 0.02$ or 2 per cent; $C = 6000$
Find the optimum rotation.
By differentiating the revenue functions in terms of t, we get:

$$\frac{dR_t}{dt} = \frac{1}{2}4000(1+t)^{-\frac{1}{2}}$$

$$= 2000(1+t)^{-\frac{1}{2}}$$

$$= \frac{2000}{(1+t)^{\frac{1}{2}}}$$

By using (6.5)

$$0.02 = \frac{2000/(1+t)^{\frac{1}{2}}}{4000(1+t)^{\frac{1}{2}}}$$

$$0.02 = \frac{1/(1+t)^{\frac{1}{2}}}{2(1+t)^{\frac{1}{2}}}$$

$$0.02 = \frac{1}{2(1+t)}$$

$$1 = (0.02)[2(1+t)]$$
$$1 = 0.04 + 0.04t$$
$$0.04t = 0.96$$
$$t = 24 \text{ years}$$

Problems with the single-rotation model

The single-rotation period is limited to a case in which the land is available in unlimited supply and thus the land rent is zero. In other words, one planting cycle only involves cash receipts other than land rent. If the land rent is not zero, then the single-rotation model, at least in the form described above, will be a misleading one. Suppose that the land is not in unlimited supply and the forest owner has the option of cutting the stand and letting out the land. Therefore, in this more general situation, the extra cost from lengthening the rotation by one period is equal to the rental value of the land which would have been paid during this period. Equation (6.5) does not capture this argument. Many well known economists as well as foresters, such as Von Thunen (1926), Hotelling (1925), and Fisher (1930) erred in their analyses when they generalised the single-rotation solution, which is indeed a very limited case. For an excellent review see Samuelson (1976).

Now, let us approach the problem from a quite different angle. It has been argued by some economists, such as Boulding (1935) and Goundrey (1960), that the internal rate of return method, which yields zero net present value (explained in Chapter 5, pp. 102–105) gives the best solution to the rotation problem in forestry economics. Instead of finding the time period that maximises the net present value, let us find a time period that maximises the internal rate of return. By rewriting equation (6.2) we get:

$$\text{NPV} = 0 = R_t e^{-xt} - C \tag{6.6}$$

Note that x is the internal rate of return, which is unknown. Our problem now is to find the time period in (6.6) that maximises the internal rate of return, which is x. To do this, we first rearrange (6.6):

$$C = R_t e^{-xt}$$

then take the natural logarithm of both sides:

$$\ln C = \ln R_t^{-xt}$$

and solve it for x:

$$\mathrm{x}t = \ln R_t - \ln C,$$

$$x = \frac{1}{t} \ln\left(\frac{R_t}{C}\right) \tag{6.7}$$

By using the revenue function in example 6.7, which is $R_t = 4000\sqrt{1+t}$, and the cost function, which is $C = 6000$, let us find which time period yields a maximum x in (6.7). For this let us try various values for t, starting from 10.

For t = *10*:

$$x = \frac{1}{10}\ln\left(\frac{4000\sqrt{1+10}}{6000}\right)$$

$$= \frac{1}{10}\ln(2.21)$$

$$= 0.079 \text{ or } 7.9 \text{ per cent.}$$

For t = *5*:

$$x = \frac{1}{5}\ln\left(\frac{4000\sqrt{1+5}}{6000}\right)$$

$$= \frac{1}{5}\ln(1.63)$$

$$= 0.098 \text{ or } 9.8 \text{ per cent}$$

For t = *4*:

$$x = \frac{1}{4}\ln\left(\frac{4000\sqrt{1+4}}{6000}\right)$$

$$= \frac{1}{4}\ln(1.49)$$

$$= 0.10 \text{ or } 10 \text{ per cent}$$

For t = *3*:

$$x = \frac{1}{3}\ln\left(\frac{4000\sqrt{1+3}}{6000}\right)$$

$$= \frac{1}{3}\ln(1.33)$$

$$= 0.095 \text{ or } 9.5 \text{ per cent}$$

The optimum period that maximises the internal rate of return is 4 years. The maximised internal rate of return, which is

10 per cent, is well above the market rate of interest, which is 2 per cent in example 6.1. However, this method of finding the optimum solution is based on an assumption that is unwarranted. The internal rate of return criterion implicitly argues that funds provided in the project can continuously be reinvested in projects with *proportional expansion of scale.* To make it more clear, in our example we start out with an investment of £4000. After four years, which is the prescribed solution to the rotation problem, the investor would re-invest:

$$£4000\ e^{(0.1)(4)} = £5969$$

and so on. As long as the internal rate is greater than the opportunity cost of capital (2 per cent in example 6.1) the present value of an infinite replication of a proportionately growing investment is infinite. Obviously, something is not quite right.

Multiple rotation with constant scale replication

The correct solution to the optimal rotation problem is to assume that the project can be replicated indefinitely at a *constant scale.* In this, once the trees are harvested, the same acreage is replanted at a constant scale over and over again. In other words, the same acreage is devoted to the afforestation project from now to eternity. This will yield a net present value formula which is:

$$\begin{aligned} \text{NPV} = {} & -C + (R_t - C)e^{-rt} + (R_t - C)e^{-2rt} \\ & + (R_t - C)e^{-3rt} + \ldots \end{aligned} \tag{6.8}$$

The first term on the right-hand side of the equation is the initial planting cost which takes place at present, year zero, for which the discount factor is one. The second term represents the discounted value of the first harvest and replanting, which take place at a future date. The third term is the second harvest and replanting, which take place at a time which is twice as long as the first one and thus it is discounted at e^{-2rt}, and so on.

Multiply both sides of equation (6.8) by e^{-rt}. This gives:

$$\text{NPV}e^{-rt} = -Ce^{-rt} + (R_t - C)e^{-rt} + \ldots + (R_t - C)e^{-(n+1)rt} \quad (6.9)$$

Subtract (6.9) from (6.8):

$$\text{NPV} - \text{NPV}e^{-rt} = -C + Ce^{-rt} + (R_t - C)e^{-rt} - (R_t - C)e^{-2rt} + \ldots - (R_t - C)e^{-(n+1)rt}$$

All the terms on the right-hand side cancel out except $-C$; Ce^{-rt}; $(R_t - C)e^{-rt}$ and $(R_t - C)e^{-(n+1)rt}$. Since n is a very large number, then $(R_t - C)e^{-(n+1)rt}$ becomes zero. Then we have

$$\text{NPV}(1 - e^{-rt}) = -C(1 - e^{-rt}) + (R_t - C)e^{-rt}$$

then

$$\text{NPV} = \frac{-C(1 - e^{-rt})}{1 - e^{-rt}} + \frac{(R_t - C)e^{-rt}}{1 - e^{-rt}}$$

which is

$$\text{NPV} = -C + \frac{(R_t - C)e^{-rt}}{1 - e^{-rt}}$$

or

$$\text{NPV} = \frac{R_t - Ce^{-rt}}{e^{rt} - 1}$$

or

$$\text{NPV} = -C + \frac{R_t - C}{e^{rt} - 1}$$

or

$$\text{NPV} = -C + (R_t - C)(e^{rt} - 1)^{-1} \quad (6.10)$$

To find the optimum rotation we have to differentiate (6.10) with respect to time and set it equal to zero (the first order maximisation condition), that is

$$\frac{\mathrm{dNPV}}{\mathrm{d}t} = \frac{\mathrm{d}}{\mathrm{d}t}\left[-C+(R_t-C)(\mathrm{e}^{rt}-1)^{-1}\right] = 0$$

$$\frac{\mathrm{d}R_t}{\mathrm{d}t}(\mathrm{e}^{rt}-1)^{-1} + r\mathrm{e}^{rt}(R_t-C)(-1)(\mathrm{e}^{rt}-1)^{-2} = 0$$

multiply both sides by $(\mathrm{e}^{rt}-1)^2$:

$$\frac{\mathrm{d}R_t}{\mathrm{d}t}(\mathrm{e}^{rt}-1) = r\mathrm{e}^{rt}(R_t-C)$$

then:

$$\frac{\mathrm{d}R_t}{\mathrm{d}t} = \frac{r\mathrm{e}^{rt}(R_t-C)}{\mathrm{e}^{rt}-1}$$

or:

$$\frac{\mathrm{d}R_t}{\mathrm{d}t} = \frac{r(R_t-C)}{1-\mathrm{e}^{-rt}} \tag{6.11}$$

Example 6.2

The revenue and cost functions are:

$$R_t = 4000(1+t)^{\frac{1}{2}}; \quad C = 6000$$

The market rate of interest r is 2 per cent; find the optimum rotation with constant scale replication.

Substituting this in (6.11) we get:

$$\frac{2000}{(1+t)^{\frac{1}{2}}} = \frac{0.02[(4000)(1+t)^{\frac{1}{2}}-6000]}{1-\mathrm{e}^{(-0.02)(t)}}$$

By trying various values of t we get the following results:

Left-hand side of equation (6.11)	*Right-hand side of equation (6.11)*	*Difference*
t = *10*:		
$\frac{2000}{(11)^{\frac{1}{2}}}$	$\frac{0.02[(4000)(11)^{\frac{1}{2}}-6000]}{1-e^{-0.2}}$	
602.41	$\frac{145.6}{1-2.72^{-0.2}}$	
602.41	$\frac{145.6}{0.18}$	
602.41	808.89	+206.48
t = *7*:		
$\frac{2000}{(8)^{\frac{1}{2}}}$	$\frac{0.02[(4000)(8)^{\frac{1}{2}}-6000]}{1-e^{-0.14}}$	
706.71	$\frac{106.4}{0.13}$	
706.71	818.46	+111.75
t = *5*:		
$\frac{2000}{(6)^{\frac{1}{2}}}$	$\frac{0.02[(4000)(6)^{\frac{1}{2}}-6000]}{1-e^{-0.1}}$	
816.33	$\frac{76}{0.095}$	
816.33	800	−16.33

Left-hand side of equation (6.11)	*Right-hand side of equation (6.11)*	*Difference*
t = *4*:		
$\frac{2000}{(5)^{\frac{1}{2}}}$	$\frac{0.02[(4000)(5)^{\frac{1}{2}}-6000]}{1-e^{-0.08}}$	
892.85	$\frac{59.2}{0.08}$	
892.85	740	−152.85

So the optimum rotation must be just over 5 years because from t=5 to t=4 the difference grows wider.

FACTORS AFFECTING THE OPTIMUM ROTATION

Taxes can extend or shorten the optimal rotation period or they can have no effect on it. Of course, when the optimal rotation is altered, this in turn will alter the volume as well as the grade of timber coming on to the market. There are a number of taxes that are relevant.

A sales tax on timber

A tax levied on the revenue resulting from the sale of timber means a decrease in the net price the forest owner receives. This leads to a longer interval between planting and harvesting.

Example 6.3

Let us take the above example of multiple rotation (example 6.2). With the imposition of sales tax the revenue function would be reduced but there are no changes in the interest rate or the cost function. That is:

$$R_t = 3500(1+t)^{\frac{1}{2}}; \quad r = 0.02; \quad C = 6000$$

Let us work out the optimal rotation by using (6.11):

$$\frac{dR_t}{dt} = \frac{1750}{(1+t)^{\frac{1}{2}}} = \frac{0.02[(3500)(1+t)^{\frac{1}{2}}-6000]}{1-e^{-0.02t}}$$

By means of the iterative method in which we try various values for t to find out which one satisfies this equation, we get $t = 7$ (approximately), which is higher than the optimal rotation period in example 6.2.

Longer rotation periods mean that the trees felled will be slightly older and more voluminous.

A negative planting tax (grant aid)

Planting grants, which are subsidies or negative taxes, will reduce the initial planting costs and this in turn, all things being equal, will shorten the rotation period. To ascertain this, let us again modify example 6.2 by reducing the planting and replanting costs by 1000.

Example 6.4

$R_t = 4000(1+t)^{\frac{1}{2}}$; $r = 0.02$; $C = 5000$

By using equation (6.11)

$$\frac{dR_t}{dt} = \frac{2000}{(1+t)^{\frac{1}{2}}} = \frac{0.02[(4000)(1+t)^{\frac{1}{2}}-5000]}{1-e^{-0.02t}}$$

The figure that satisfies this equation is $t = 3$ (approximately).

It was explained in Chapter 5 that afforestation projects in the United Kingdom are heavily subsidised by governments to encourage private entrepreneurs to take up such projects. Note that the improved grant would affect optimum rotation on the new as well as the existing plantations. The subsidy would affect the existing plantations via the reduction in the cost of replanting.

A site utilisation tax

The government may impose a tax per hectare each time land is brought into forestry use. This would be equivalent to an increase in the establishment cost, which would increase the profit-maximising rotation period. As a result the trees cut down would be older and more voluminous.

A licence fee for afforestation

A licence fee implemented on a per hectare basis would have a similar effect to a site utilisation tax as it would, effectively, increase set-up costs. For example, the government may enforce a licensing requirement, at a cost, on foresters. This would be on the basis of per hectare or per year. If the licence fee were levied per year rather than per hectare, then the optimal rotation interval would be unaffected.

A profit tax

A fixed percentage tax on profits resulting from the sale of timber would not change the optimal rotation interval because it would have no bearing on the parameters that determine the optimum interval.

Changes in interest rates

Since the interest rate is a dominant parameter in equation (6.11), any change in this will alter the final result. It must be intuitively obvious that a high interest rate will shorten the optimal rotation point. Going back to the argument regarding a single rotation, in Figure 6.2 a high interest rate would mean a steeper line for the opportunity cost of capital and hence the tangency point will be nearer to the origin. Of course, a high interest rate will also make investors reluctant to tie up their money for a longer period of time in forestry projects.

Example 6.5

The revenue and cost functions are:

$$R_t = 4000\,(1+t)^{\frac{1}{2}}$$
$$C = 6000$$
$$r = 5 \text{ per cent}$$

Find the optimum rotation.

$$\frac{dR_t}{dt} = \frac{2000}{(1+t)^{\frac{1}{2}}}$$

By using equation (6.11) we get:

$$\frac{2000}{(1+t)^{\frac{1}{2}}} = \frac{0.05\,[(4000)(1+t)^{\frac{1}{2}} - 6000]}{1 - e^{-0.05t}}$$

Again, by using the iterative method in which we try various values for t to find out which one yields the desired solution, we get $t = 4$ (years) as opposed to $t = 5$ (years) in example 6.2 above.

Therefore, other things being equal, the higher the interest rate the shorter the rotation. This conclusion will become important in the following section where the optimum rotation in the public sector is analysed.

OPTIMAL ROTATION IN PUBLIC SECTOR FORESTRY

The objective of a private entrepreneur is to maximise the net cash benefits arising from his project. The analysis which has so far been carried out in this chapter enables the private investor to maximise his profits by choosing the most advantageous rotation period. The parameters of the equation which he needs are market rate of interest, planting and replanting costs, and the cash revenue arising from the sale of timber. However, in public sector economics the objective of a government agency which is investing in a public project is always assumed to be the maximisation of net social benefits accruing to the community from that project. Apart from the economic gain there could be other objectives, e.g. cultural, political and military uses.

As explained in Chapter 1, the military problems encountered during the two World Wars contributed immensely to the establishment of forestry authorities in the British Isles, and in the case of Ireland the expansion of an already existing body. It would therefore be quite misleading to regard forests as purely economic and commercial assets.

Political, military and cultural objectives aimed at by the government are of course crucial. However, the economic objective, that is, the maximisation of net cash benefits to the community, should not be pushed aside as trivial. The overall objective of the government must be the maximisation of the sum of total utility over time. That is:

$$\max_{C(t)} \int_0^\infty U[C(t)](e^{-st})^* \, dt \tag{6.12}$$

where

$C(t)$ = real net benefit arising from forestry at time t
U = social utility
s = social discount rate
$(e^{-st})^*$ = discount factor by means of modified discounting

In this it is assumed that the decision maker treats all generations in an equitable manner and for this uses a social interest rate, as opposed to market rate, and discounting is conducted in a just manner as suggested in Chapter 4. In other words at, say, a 10 per cent discount rate the discount factor $(e^{-st})^*$ corresponds to the MDM curve in Figure 6.4; not the NPV curve in the same figure. Also note that the utility function which is to be maximised should include not only the benefits arising from the production of timber, but also the environmental, recreational, cultural and political (i.e. through the feeling of military security) advantages.

The social discount rate

The discount rate which is relevant in the economic analysis of public sector forestry is, of course, the social interest rate as opposed to the market rate. Now the question must be: what are the criteria by which the social discount rate is arrived at? There is a very long-winded and inconclusive argument in the economics literature about the choice of this rate. There are, basically, three ideas: the government borrowing rate, the social opportunity cost rate and the social time preference rate.

Figure 6.4: Discount factor curves by means of the sum of discounted consumption flows (MDM) and the net present value (NPV) criteria at 10 and 5 per cent discount rates

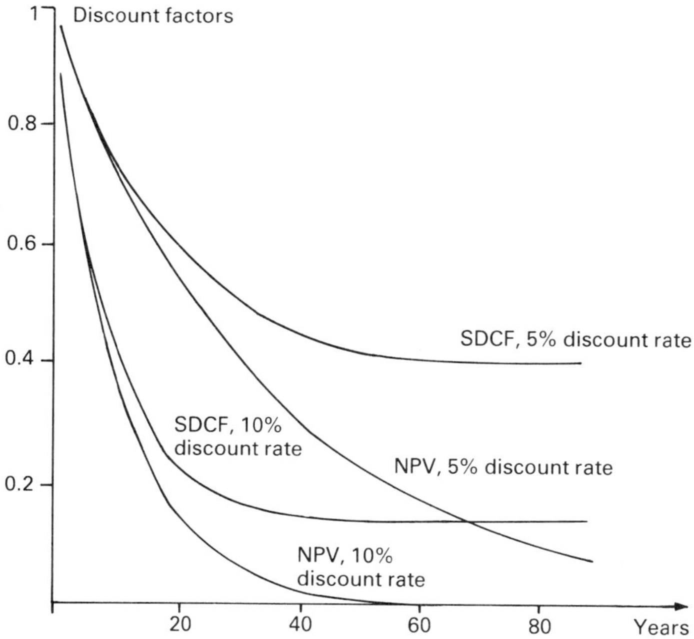

The government borrowing rate

The use of the long-term government borrowing rate as the appropriate social discount rate in public sector economics was the earliest practice. During the 1930s in the US, when cost–benefit analysis was first used to evaluate the worth of various land-based projects, the discount rates employed by the appraisal agencies were very much in line with the long-term government borrowing rate. There were two justifications for using such a rate. First, some economists related the government borrowing rate to the risk-free rate of return in the economy. Their argument was that since public sector projects can be considered as risk-free, because of a large portfolio situation, then only a risk-free interest rate was suitable in their evaluation. Secondly, the government borrowing rate represented the cost of capital used in the construction of public sector projects. However, under close scrutiny these arguments proved to be false and thus this school has lost credibility.

It is not correct to claim that government bonds are risk-free. In fact, they are subject to two types of risk. First, variations in the purchasing power of money affect all real interest rates including the ones promised by bonds. Secondly, the market value of a bond is affected by fluctuations in the interest rate. For example, take a perpetual bond of £1000 issued at a fixed interest rate of 5 per cent at the time when the ruling market rate was also 5 per cent. If the market rate goes up to 10 per cent, then nobody will pay £1000 for this bond as one could earn twice as much by putting £1000 into a bank account. Therefore, doubling of the market interest rate would halve the market value of the bond. As for the second alleged merit of the government bond rate as being the cost of capital used to finance public projects, this too is a dubious argument. In reality, governments raise only a small proportion of their revenue by borrowing, the bulk of their money coming from taxation. Therefore, nowadays most economists do not advocate the use of government borrowing rate in public sector projects.

The social opportunity cost rate

A good number of economists argue that, since capital funds are limited, a public sector investment project will displace some other projects in the economy. Therefore, in evaluation of any public sector project the discount rate used must be the one that reflects the social opportunity cost of capital. This rate measures the value to society of the next best alternative investment in which funds might otherwise have been employed. Generally, the next best alternative investment is thought to be in the private sector (Krutilla and Eckstein, 1958, Hirshleifer *et al.*,

1960, Kuhn, 1962, Joint Economic Committee, US Congress, 1968).

There are some serious problems associated with this approach. First, the stock market's view of a rate of return on capital can be very different from society's view of profitability. In private profitability accounts, external costs such as noise, pollution and congestion which a project may generate are not considered in calculations. Another point is that private profits may be quite high, not as a result of an efficient operation but as a result of market imperfections such as monopoly, cartel, oligopoly, etc., all of which work against the public interest. Therefore, most economists believe that the private profits and resulting rates of return on capital require a substantial social adjustment before they can be used in public sector project evaluation.

The social time preference rate

Another school of economists argues that the only suitable rate of discount to be used in public sector investment appraisal is the social time preference rate, STPR, which is also called the consumption rate of interest, CRI, or the social rate of interest, SRI. It is the rate by which the decision maker discounts, on behalf of society, the costs and benefits which arise from public sector projects. The rationale for their argument is quite simple. The purpose behind investment decisions is to increase future consumption, which involves a sacrifice on present consumption. In other words, here consumption is the name of the game and thus the relevant rate of discount is the consumption rate of interest.

What constitutes a social time preference rate? There is a long-winded argument behind this question. There seem to be two rational factors which may justify discounting: diminishing marginal utility of consumption, and risk of death. As regards the former it is argued that in most societies the standard of living enjoyed by individuals is generally rising. As the income of a person increases steadily, the satisfaction gained also increases but at a slower rate. That is, each addition to his income yields a successively smaller increase to his economic welfare. Therefore £1 now should mean more to an individual than £1 later on. As regards the latter, it is argued that the risk of death is a powerful factor which affects an individual when he is making intertemporal consumption decisions. His preference for present consumption over future consumption of the same magnitude is perfectly sensible since he may not be alive to consume it in the future.

For derivation of social time preference rates for the UK, the US and Canada, see Appendix III. The figures shown there do not seem to be very different from the average long-term market rates of interest which have prevailed in these countries. Therefore, the real issue in finding the optimum rotation for public sector forestry seems not to be the difference between the private and social interest rates in magnitude, although some may dispute this, but the way in which the discounting is conducted. As explained in Chapter 4, modified discounting yields a much more moderate discounting pattern than the conventional methods, a case which can also be seen in Figure 6.4.

To conclude, in public sector forestry the optimum rotation period will be much longer than in private sector forestry, mainly because of the modified discounting method.

Foresters traditionally defended a rotation period, the maximum sustainable yield, which yields maximum wood output. Some economists such as Samuelson (1976) argued that this was a wrong criterion and thus should not be used in determining the optimum rotation period. His argument is that the optimal solution to the rotation problem requires a proper account of two economic factors, the rental value of the land on which the trees are grown, and the interest rate, which is the cost of capital. With the inclusion of these the economic solution to the rotation problem will be shorter than the one advocated by foresters. In his paper Samuelson draws up a list of foresters as well as economists who, over the years, prescribed misleading results to the optimum rotation problem. He argues that some of these results, such as the maximum sustainable yield, are very wrong, some just wrong.

When we use the modified discounting method to find the optimum rotation for public sector afforestation projects, the outcome will be much closer to the maximum sustainable yield than the criterion advocated by Samuelson and others. I have argued on many occasions that the use of ordinary discounting in economic analyses of public sector projects is both conceptually and morally wrong. So far, almost all articles published in cost-benefit analysis have made explicit or implicit reference to ordinary discounting and it has always surprised me how an issue like this, which is so crucial to economics, has escaped proper scrutiny for so long.

Appendix I: Discounted cash flows discount factors

Year	Percentage 4.0	5.0	6.0	7.0	8.0
1	0.9615	0.9524	0.9434	0.9346	0.9259
2	0.9246	0.9070	0.8900	0.8734	0.8573
3	0.8890	0.8638	0.8396	0.8163	0.7938
4	0.8548	0.8227	0.7921	0.7629	0.7350
5	0.8219	0.7835	0.7473	0.7130	0.6806
6	0.7903	0.7462	0.7050	0.6663	0.6302
7	0.7599	0.7107	0.6651	0.6228	0.5835
8	0.7307	0.6768	0.6274	0.5820	0.5403
9	0.7026	0.6446	0.5919	0.5439	0.5002
10	0.6756	0.6139	0.5584	0.5083	0.4632
11	0.6496	0.5847	0.5268	0.4751	0.4289
12	0.6246	0.5568	0.4970	0.4440	0.3971
13	0.6006	0.5303	0.4688	0.4150	0.3677
14	0.5775	0.5051	0.4423	0.3878	0.3405
15	0.5553	0.4810	0.4173	0.3624	0.3152
16	0.5339	0.4581	0.3936	0.3387	0.2919
17	0.5134	0.4363	0.3714	0.3166	0.2703
18	0.4936	0.4155	0.3503	0.2959	0.2502
19	0.4746	0.3957	0.3305	0.2765	0.2317
20	0.4564	0.3769	0.3118	0.2584	0.2145
21	0.4388	0.3589	0.2942	0.2415	0.1987
22	0.4219	0.3419	0.2775	0.2257	0.1839
23	0.4057	0.3256	0.2618	0.2109	0.1703
24	0.3901	0.3101	0.2470	0.1971	0.1577
25	0.3751	0.2953	0.2330	0.1842	0.1460
26	0.3607	0.2812	0.2198	0.1722	0.1352
27	0.3468	0.2678	0.2074	0.1609	0.1252
28	0.3335	0.2551	0.1956	0.1504	0.1159
29	0.3207	0.2429	0.1846	0.1406	0.1073
30	0.3083	0.2314	0.1741	0.1314	0.0994
31	0.2965	0.2204	0.1643	0.1228	0.0920
32	0.2851	0.2099	0.1550	0.1147	0.0852
33	0.2741	0.1999	0.1462	0.1072	0.0789
34	0.2636	0.1904	0.1379	0.1002	0.0730
35	0.2534	0.1813	0.1301	0.0937	0.0676
36	0.2437	0.1727	0.1227	0.0875	0.0626
37	0.2343	0.1644	0.1158	0.0818	0.0580
38	0.2253	0.1566	0.1092	0.0765	0.0537
39	0.2166	0.1491	0.1031	0.0715	0.0497
40	0.2083	0.1420	0.0972	0.0668	0.0460
41	0.2003	0.1353	0.0917	0.0624	0.0426
42	0.1926	0.1288	0.0865	0.0583	0.0395
43	0.1852	0.1227	0.0816	0.0545	0.0365
44	0.1780	0.1169	0.0770	0.0509	0.0338
45	0.1712	0.1113	0.0727	0.0476	0.0313
46	0.1646	0.1060	0.0685	0.0445	0.0290
47	0.1583	0.1009	0.0647	0.0416	0.0269
48	0.1522	0.0961	0.0610	0.0389	0.0249
49	0.1463	0.0916	0.0575	0.0363	0.0230
50	0.1407	0.0872	0.0543	0.0339	0.0213

Year	Percentage				
	9.0	10.0	11.0	12.0	15.0
1	0.9174	0.9091	0.9009	0.8929	0.8696
2	0.8417	0.8264	0.8116	0.7972	0.7561
3	0.7722	0.7513	0.7312	0.7118	0.6575
4	0.7084	0.6830	0.6587	0.6355	0.5718
5	0.6499	0.6209	0.5935	0.5674	0.4972
6	0.5963	0.5645	0.5346	0.5066	0.4323
7	0.5470	0.5132	0.4817	0.4523	0.3759
8	0.5019	0.4665	0.4339	0.4039	0.3269
9	0.4604	0.4241	0.3909	0.3606	0.2843
10	0.4224	0.3855	0.3522	0.3220	0.2472
11	0.3875	0.3505	0.3173	0.2875	0.2149
12	0.3555	0.3186	0.2858	0.2567	0.1869
13	0.3262	0.2897	0.2575	0.2292	0.1625
14	0.2992	0.2633	0.2320	0.2046	0.1413
15	0.2745	0.2394	0.2090	0.1827	0.1229
16	0.2519	0.2176	0.1883	0.1631	0.1069
17	0.2311	0.1978	0.1696	0.1456	0.0929
18	0.2120	0.1799	0.1528	0.1300	0.0808
19	0.1945	0.1635	0.1377	0.1161	0.0703
20	0.1784	0.1486	0.1240	0.1037	0.0611
21	0.1637	0.1351	0.1117	0.0926	0.0531
22	0.1502	0.1228	0.1007	0.0826	0.0462
23	0.1378	0.1117	0.0907	0.0738	0.0402
24	0.1264	0.1015	0.0817	0.0659	0.0349
25	0.1160	0.0923	0.0736	0.0588	0.0304
26	0.1064	0.0839	0.0663	0.0525	0.0264
27	0.0976	0.0763	0.0597	0.0469	0.0230
28	0.0895	0.0693	0.0538	0.0419	0.0200
29	0.0822	0.0630	0.0485	0.0374	0.0174
30	0.0754	0.0573	0.0437	0.0334	0.0151
31	0.0691	0.0521	0.0394	0.0298	0.0131
32	0.0634	0.0474	0.0355	0.0266	0.0114
33	0.0582	0.0431	0.0319	0.0238	0.0099
34	0.0534	0.0391	0.0288	0.0212	0.0086
35	0.0490	0.0356	0.0259	0.0189	0.0075
36	0.0449	0.0323	0.0234	0.0169	0.0065
37	0.0412	0.0294	0.0210	0.0151	0.0057
38	0.0378	0.0267	0.0190	0.0135	0.0049
39	0.0347	0.0243	0.0171	0.0120	0.0043
40	0.0318	0.0221	0.0154	0.0107	0.0037
41	0.0292	0.0201	0.0139	0.0096	0.0032
42	0.0268	0.0183	0.0125	0.0086	0.0028
43	0.0246	0.0166	0.0112	0.0076	0.0025
44	0.0226	0.0151	0.0101	0.0068	0.0021
45	0.0207	0.0137	0.0091	0.0061	0.0019
46	0.0190	0.0125	0.0082	0.0054	0.0016
47	0.0174	0.0113	0.0074	0.0049	0.0014
48	0.0160	0.0103	0.0067	0.0043	0.0012
49	0.0147	0.0094	0.0060	0.0039	0.0011
50	0.0134	0.0085	0.0054	0.0035	0.0009

Appendix II: Discount factors for the United Kingdom on the basis of MDM

	Discount rate, percentage							
Year (*t*)	1	2	3	4	5	6	7	8
1	0.99010	0.98039	0.97087	0.96154	0.95238	0.94340	0.93458	0.92593
2	0.98043	0.96143	0.94298	0.92506	0.90765	0.89073	0.87428	0.85828
3	0.97099	0.94310	0.91628	0.89048	0.86564	0.84173	0.81870	0.79631
4	0.96178	0.92538	0.89072	0.85769	0.82620	0.79616	0.76749	0.74013
5	0.95278	0.90826	0.86626	0.82661	0.78917	0.75378	0.72032	0.68866
6	0.94401	0.89172	0.84286	0.79717	0.75441	0.71438	0.67687	0.64170
7	0.93545	0.87574	0.82047	0.76927	0.72180	0.67776	0.63686	0.59886
8	0.92710	0.86031	0.79906	0.74284	0.69120	0.64372	0.60003	0.55978
9	0.91896	0.84541	0.77859	0.71782	0.66250	0.61210	0.56612	0.52415
10	0.91102	0.83102	0.75902	0.69413	0.63559	0.58273	0.53493	0.49166
11	0.90328	0.81714	0.74031	0.67171	0.61036	0.55545	0.50622	0.46205
12	0.89575	0.80375	0.72244	0.65049	0.58672	0.53012	0.47983	0.43507
13	0.88840	0.79083	0.70537	0.63041	0.56456	0.50662	0.45555	0.41049
14	0.88125	0.77837	0.68907	0.61143	0.54380	0.48481	0.43324	0.38810
15	0.87429	0.76636	0.67351	0.59348	0.52437	0.46457	0.41274	0.36772
16	0.86752	0.75478	0.65865	0.57651	0.50617	0.44580	0.39390	0.34917
17	0.86092	0.74363	0.64448	0.56047	0.48913	0.42841	0.37659	0.33228
18	0.85451	0.73289	0.63096	0.54533	0.47320	0.41228	0.36071	0.31693
19	0.84827	0.72254	0.61807	0.53102	0.45829	0.39734	0.34612	0.30296
20	0.84221	0.71259	0.60579	0.51752	0.44435	0.38350	0.33274	0.29026
21	0.83632	0.70301	0.59408	0.50477	0.43132	0.37069	0.32047	0.27872
22	0.83060	0.69379	0.58292	0.49275	0.41914	0.35883	0.30921	0.26824
23	0.82504	0.68493	0.57230	0.48141	0.40777	0.34785	0.29890	0.25872
24	0.81965	0.67642	0.56220	0.47073	0.39715	0.33770	0.28945	0.25008
25	0.81441	0.66823	0.55258	0.46065	0.38724	0.32832	0.28079	0.24224
26	0.80934	0.66037	0.54343	0.45117	0.37799	0.31965	0.27286	0.23513
27	0.80441	0.65283	0.53474	0.44224	0.36937	0.31164	0.26561	0.22868
28	0.79964	0.64559	0.52647	0.43383	0.36133	0.30424	0.25898	0.22284
29	0.79503	0.63865	0.51863	0.42592	0.35385	0.29741	0.25291	0.21755
30	0.79055	0.63199	0.51118	0.41849	0.34687	0.29112	0.24737	0.21275
31	0.78623	0.62562	0.50411	0.41150	0.34038	0.28531	0.24231	0.20842
32	0.78204	0.61951	0.49741	0.40494	0.33435	0.27996	0.23768	0.20450
33	0.77800	0.61367	0.49105	0.39879	0.32873	0.27504	0.23347	0.20095
34	0.77409	0.60808	0.48504	0.39301	0.32352	0.27050	0.22962	0.19775
35	0.77032	0.60274	0.47934	0.38759	0.31867	0.26633	0.22612	0.19486
36	0.76668	0.59763	0.47395	0.38252	0.31418	0.26250	0.22293	0.19225
37	0.76317	0.59276	0.46886	0.37777	0.31001	0.25898	0.22003	0.18990
38	0.75980	0.58811	0.46405	0.37333	0.30615	0.25575	0.21739	0.18778
39	0.75654	0.58368	0.45950	0.36917	0.30258	0.25278	0.21499	0.18587
40	0.75341	0.57946	0.45522	0.36529	0.29927	0.25006	0.21281	0.18416
41	0.75041	0.57545	0.45119	0.36167	0.29621	0.24758	0.21084	0.18262
42	0.74752	0.57163	0.44739	0.35829	0.29339	0.24530	0.20905	0.18123
43	0.74475	0.56801	0.44381	0.35515	0.29078	0.24322	0.20743	0.17999
44	0.74210	0.56457	0.44045	0.35222	0.28838	0.24132	0.20596	0.17888
45	0.73956	0.56131	0.43730	0.34950	0.28617	0.23959	0.20464	0.17788
46	0.73714	0.55823	0.43435	0.34698	0.28414	0.23801	0.20344	0.17699
47	0.73482	0.55531	0.43158	0.34463	0.28227	0.23658	0.20237	0.17620
48	0.73261	0.55256	0.42900	0.34247	0.28056	0.23527	0.20140	0.17549
49	0.73051	0.54996	0.42658	0.34046	0.27899	0.23409	0.20053	0.17486

Year (*t*)	1	2	3	4	5	6	7	8
50	0.72851	0.54752	0.42433	0.33861	0.27756	0.23302	0.19975	0.17430
51	0.72661	0.54522	0.42224	0.33691	0.27625	0.23205	0.19905	0.17380
52	0.72481	0.54307	0.42029	0.33534	0.27506	0.23118	0.19842	0.17336
53	0.72312	0.54106	0.41849	0.33390	0.27397	0.23039	0.19786	0.17297
54	0.72152	0.53917	0.41683	0.33258	0.27299	0.22969	0.19737	0.17263
55	0.72001	0.53742	0.41529	0.33138	0.27210	0.22905	0.19693	0.17233
56	0.71860	0.53580	0.41388	0.33028	0.27130	0.22849	0.19654	0.17206
57	0.71728	0.53429	0.41258	0.32928	0.27058	0.22798	0.19619	0.17183
58	0.71605	0.53290	0.41140	0.32838	0.26993	0.22753	0.19589	0.17163
59	0.71490	0.53162	0.41032	0.32757	0.26935	0.22714	0.19562	0.17145
60	0.71385	0.53045	0.40934	0.32684	0.26884	0.22679	0.19539	0.17130
61	0.71288	0.52939	0.40846	0.32619	0.26839	0.22648	0.19519	0.17117
62	0.71199	0.52843	0.40767	0.32561	0.26799	0.22622	0.19502	0.17106
63	0.71119	0.52756	0.40697	0.32510	0.26764	0.22599	0.19487	0.17096
64	0.71046	0.52679	0.40635	0.32466	0.26734	0.22579	0.19474	0.17088
65	0.70982	0.52611	0.40581	0.32427	0.26708	0.22562	0.19464	0.17082
66	0.70925	0.52551	0.40534	0.32394	0.26686	0.22548	0.19455	0.17076
67	0.70875	0.52501	0.40495	0.32366	0.26668	0.22536	0.19447	0.17072
68	0.70834	0.52458	0.40462	0.32344	0.26653	0.22527	0.19442	0.17068
69	0.70799	0.52423	0.40435	0.32325	0.26641	0.22520	0.19437	0.17066
70	0.70772	0.52396	0.40414	0.32311	0.26632	0.22514	0.19434	0.17064
71	0.70752	0.52375	0.40399	0.32301	0.26625	0.22510	0.19431	0.17062
72	0.70738	0.52362	0.40389	0.32295	0.26621	0.22508	0.19430	0.17062
73	0.70732	0.52356	0.40385	0.32291	0.26619	0.22507	0.19429	0.17061

	Discount rate, percentage						
Year (*t*)	9	10	11	12	13	14	15
1	0.91743	0.90909	0.90090	0.89286	0.88496	0.87719	0.86957
2	0.84272	0.82758	0.81285	0.79850	0.78454	0.77094	0.75770
3	0.77513	0.75451	0.73462	0.71543	0.69691	0.67904	0.66177
4	0.71399	0.68901	0.66514	0.64230	0.62046	0.59955	0.57953
5	0.65870	0.63032	0.60343	0.57794	0.55377	0.53082	0.50904
6	0.60871	0.57774	0.54865	0.52131	0.49560	0.47141	0.44864
7	0.56352	0.53064	0.50002	0.47149	0.44489	0.42006	0.39688
8	0.52269	0.48846	0.45687	0.42767	0.40067	0.37569	0.35255
9	0.48579	0.45070	0.41858	0.38914	0.36214	0.33735	0.31458
10	0.45246	0.41690	0.38461	0.35527	0.32857	0.30425	0.28207
11	0.42236	0.38665	0.35449	0.32550	0.29932	0.27566	0.25425
12	0.39518	0.35959	0.32779	0.29934	0.27385	0.25098	0.23044
13	0.37065	0.33539	0.30412	0.27636	0.25167	0.22968	0.21007
14	0.34852	0.31374	0.28314	0.25617	0.23236	0.21130	0.19264
15	0.32855	0.29439	0.26456	0.23846	0.21556	0.19545	0.17774
16	0.31054	0.27710	0.24811	0.22290	0.20095	0.18178	0.16501
17	0.29430	0.26166	0.23354	0.20926	0.18824	0.17000	0.15412
18	0.27966	0.24786	0.22064	0.19729	0.17719	0.15984	0.14483
19	0.26647	0.23554	0.20923	0.18679	0.16758	0.15109	0.13688
20	0.25459	0.22454	0.19914	0.17759	0.15924	0.14355	0.13010
21	0.24390	0.21473	0.19021	0.16952	0.15199	0.13707	0.12432

Year (t)	Discount rate, percentage 9	10	11	12	13	14	15
22	0.23427	0.20598	0.18233	0.16246	0.14569	0.13148	0.11938
23	0.22561	0.19818	0.17536	0.15627	0.14023	0.12668	0.11517
24	0.21781	0.19123	0.16920	0.15086	0.13549	0.12255	0.11158
25	0.21081	0.18503	0.16377	0.14612	0.13138	0.11900	0.10852
26	0.20451	0.17951	0.15897	0.14197	0.12782	0.11595	0.10592
27	0.19886	0.17460	0.15474	0.13835	0.12473	0.11333	0.10370
28	0.19378	0.17023	0.15101	0.13519	0.12206	0.11108	0.10181
29	0.18922	0.16635	0.14772	0.13242	0.11974	0.10914	0.10021
30	0.18513	0.16289	0.14482	0.13001	0.11774	0.10749	0.09884
31	0.18149	0.15982	0.14227	0.12790	0.11601	0.10607	0.09768
32	0.17818	0.15710	0.14003	0.12606	0.11451	0.10485	0.09670
33	0.17524	0.15468	0.13806	0.12446	0.11322	0.10381	0.09586
34	0.17261	0.15254	0.13632	0.12307	0.11210	0.10292	0.09515
35	0.17025	0.15063	0.13480	0.12185	0.11113	0.10216	0.09455
36	0.16815	0.14895	0.13346	0.12080	0.11030	0.10150	0.09404
37	0.16626	0.14746	0.13229	0.11988	0.10959	0.10095	0.09361
38	0.16459	0.14614	0.13126	0.11908	0.10897	0.10047	0.09324
39	0.16309	0.14498	0.13036	0.11839	0.10844	0.10007	0.09293
40	0.16175	0.14395	0.12957	0.11779	0.10799	0.09972	0.09267
41	0.16057	0.14304	0.12888	0.11727	0.10759	0.09943	0.09245
42	0.15951	0.14224	0.12828	0.11682	0.10726	0.09918	0.09227
43	0.15857	0.14153	0.12775	0.11643	0.10697	0.09897	0.09211
44	0.15773	0.14091	0.12730	0.11609	0.10672	0.09879	0.09198
45	0.15699	0.14037	0.12690	0.11580	0.10651	0.09863	0.09187
46	0.15634	0.13989	0.12655	0.11555	0.10633	0.09850	0.09177
47	0.15576	0.13947	0.12625	0.11533	0.10618	0.09839	0.09170
48	0.15525	0.13910	0.12599	0.11515	0.10605	0.09830	0.09163
49	0.15480	0.13878	0.12576	0.11499	0.10593	0.09822	0.09158
50	0.15440	0.13850	0.12557	0.11485	0.10584	0.09816	0.09153
51	0.15405	0.13826	0.12540	0.11473	0.10576	0.09810	0.09149
52	0.15374	0.13805	0.12525	0.11463	0.10569	0.9806	0.09146
53	0.15347	0.13786	0.12513	0.11455	0.10563	0.09802	0.09144
54	0.15324	0.13770	0.12502	0.11448	0.10558	0.09798	0.09141
55	0.15303	0.13757	0.12493	0.11442	0.10554	0.09796	0.09140
56	0.15285	0.13745	0.12485	0.11436	0.10551	0.09794	0.09138
57	0.15270	0.13735	0.12478	0.11432	0.10548	0.09792	0.09137
58	0.15257	0.13726	0.12472	0.11428	0.10546	0.09790	0.09136
59	0.15245	0.13718	0.12468	0.11425	0.10544	0.09789	0.09135
60	0.15235	0.13712	0.12464	0.11423	0.10542	0.09788	0.09134
61	0.15227	0.13707	0.12460	0.11421	0.10541	0.09789	0.09134
62	0.15220	0.13702	0.12457	0.11419	0.10540	0.09786	0.09133
63	0.15214	0.13699	0.12455	0.11417	0.10539	0.09786	0.09133
64	0.15209	0.13696	0.12453	0.11416	0.10538	0.09785	0.09133
65	0.15205	0.13693	0.12452	0.11415	0.10538	0.09785	0.09133
66	0.15202	0.13691	0.12450	0.11414	0.10537	0.09785	0.09132
67	0.15199	0.13689	0.12449	0.11414	0.10537	0.09784	0.09132
68	0.15197	0.13688	0.12449	0.11413	0.10537	0.09784	0.09132
69	0.15195	0.13687	0.12448	0.11413	0.10536	0.09784	0.09132
70	0.15194	0.13686	0.12448	0.11413	0.10536	0.09784	0.09132
71	0.15193	0.13686	0.12447	0.11413	0.10536	0.09784	0.09132
72	0.15193	0.13686	0.12447	0.11413	0.10536	0.09784	0.09132
73	0.15192	0.13686	0.12447	0.11413	0.10536	0.09784	0.09132

APPENDIX III

The author has, over the years, developed models to estimate a social time preference rate (Kula 1985, 1986a, d, 1987a). The simplest one is as follows.* Technically speaking, in a two-period analysis the STPR corresponds to the marginal rate of substitution of consumption along the social indifference curve. In Figure AIII.1 the trade-off between present and future consumption is shown along the social indifference curve II. If the community is located at a point such as S, then the STPR will be equal to the marginal rate of substitution of consumption at this point, minus one. That is:

$$S = MRSC_{0,1} - 1 \qquad \text{(AIII.1)}$$

Figure AIII.1: Marginal substitution of consumption between present and future and resulting time preference rates

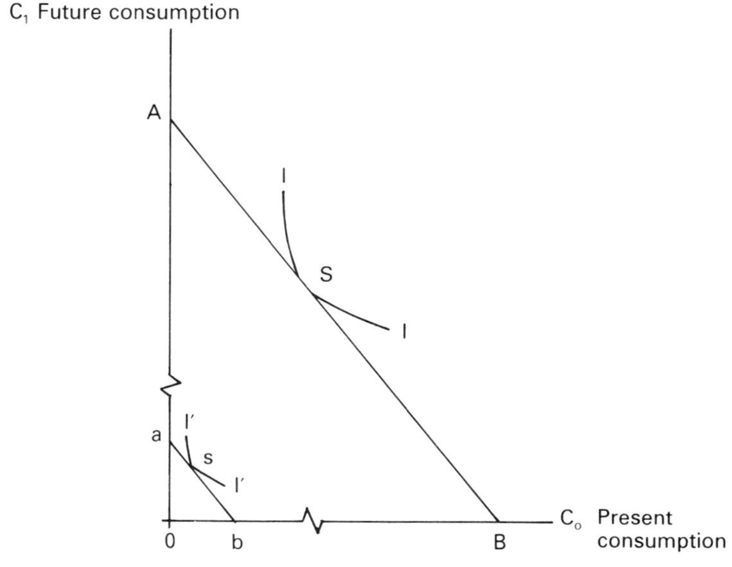

*The analysis which is carried out in this appendix may prove to be difficult for foresters, geographers and environmental scientists to follow as it requires a solid background in economics. My advice to them would be to look only at the results given at the end.

where S = social time preference rate, MRSC = marginal rate of substitution of consumption, and subscripts 0 and 1 refer to present and future respectively.

In order to arrive at an operative formula, the following argument is suggested. Let us assume a typical individual, Mr Average, who represents the community at a given point in time. He has an indifference curve, shown as I′I′ in Figure AIII.1, which is the miniaturised version of the social indifference curve. Along this we locate him at point s, where the marginal rate of substitution is the same as point S on the social indifference curve, i.e. 0A/0B = 0a/0b. Then his time preference equals the STPR. He has, say, a two-period consumption utility function, *u*, which has a constant elasticity, and its net present value is:

$$\mathrm{NPV}(u) = \frac{(A)(C_0)^{1-e}}{(1-e)} + \frac{(A)(C_1)^{1-e}}{(1-e)(1+m)} \qquad \text{(AIII.2)}$$

where:

C = real consumption with subscripts from year 0 to year 1
A = constant
e = elasticity of marginal utility of consumption
m = risk of death from one period to the next.

The marginal rate of substitution is a ratio which is arrived at by differentiating (AIII.2) by C_0 and C_1 and dividing $\mathrm{dNPV}(u)/\mathrm{d}C_0$ by $\mathrm{dNPV}(u)/\mathrm{d}C_1$, that is:

$$\frac{\mathrm{dNPV}(u)}{\mathrm{d}C_0} = (A)(C_0)^{-e}$$

$$\frac{\mathrm{dNPV}(u)}{\mathrm{d}C_1} = \frac{(A)(C_1)^{-e}}{(1+m)}$$

Then the marginal rate of substitution is:

$$\mathrm{MRSC}_{0,1} = \frac{\mathrm{dNPV}(u)}{\mathrm{d}C_0} \Big/ \frac{\mathrm{dNPV}(u)}{\mathrm{d}C_1} = (1+m)\left(\frac{C_1}{C_0}\right)^e$$

Then by definition the STPR is:

$$S = (1+m)\left(\frac{C_1}{C_0}\right)^e - 1 \quad \text{(AIII.3)}$$

The growth rate of the individual's real consumption between these two time points is:

$$g = \frac{(C_1 - C_0)}{C_0}$$

where g = growth rate of real consumption.

This yields:

$$(1+g) = \frac{C_1}{C_0}$$

Substituting this in (AIII.3) we get:

$$S = (1+m)(1+g)^e - 1 \quad \text{(AIII.4)}$$

By linear approximation (AIII.4) can be written as:

$$S = e(g) + m \quad \text{(AIII.5)}$$

Estimates of social time preference rates for the UK, US and Canada

What are the likely estimates of social time preference rates for the UK, the US and Canada on the basis of (AIII.5)? Let us take the mortality-based pure time discount rate, *m*, first. In 1985 the UK population was 55.9 million and the number of deaths was 655 thousand giving, on average, a death rate of 1.17 per cent. The crude death rates calculated in this fashion for these countries since the war are shown for some selective years in Table AIII.1. The figures for the US and Canada have remained virtually unchanged at 0.99 per cent whereas the UK's figure seems to hover around 1.17 per cent. These numbers are taken to represent the risk of death faced by the Mr Average in these countries. That is:

Country	*Mortality based pure time discount rate,* m (%)
UK	1.17
US	0.99
Canada	0.99

As for the growth rate of real consumption, g, it is obtained by fitting the equation

$$\ln(C) = a+gt \tag{AIII.6}$$

to the series in Table AIII.2 where C is per capita real consumption, a is a constant and t is the number of years between 1954 and 1976 (inclusive). The results are:

Country	*Growth rate of real consumption,* g (%)
UK	2.0
US	2.3
Canada	2.8

As for the third component parameter, that is, the elasticity of marginal utility of consumption, e, a relatively simple model which was initiated by Fisher (1927) and Frisch (1932) but more recently developed by Fellner (1967) can be used. In this model the elasticity of marginal utility of consumption is measured by the ratio of income elasticity to pure price elasticity of the food demand function, that is,

$$e = \frac{y}{\hat{p}} \tag{AIII.7}$$

Table AIII.1: Annual crude death rates (%) in the UK, US and Canada

	Years						
Countries	1946	1950	1955	1960	1965	1970	1975
UK	1.16	1.17	1.17	1.15	1.16	1.20	1.18
US	0.99	0.99	0.99	0.99	0.99	0.99	0.99
Canada	0.99	0.99	0.99	0.99	0.99	0.99	0.99

Source: Kula (1987b).

Table AIII.2: Series for equation (AIII.6)

Year (t)	*Per capita* real consumption (*C*) UK, £ (1970 prices)	US, $ (1967 prices)	Canada, $ (1967 prices)
1954	415	1810	1258
1955	431	1913	1315
1956	431	1972	1371
1957	440	1976	1371
1958	449	1945	1375
1959	466	2031	1412
1960	480	2043	1415
1961	487	2036	1402
1962	493	2100	1446
1963	512	2152	1488
1964	526	2235	1539
1965	529	2340	1594
1966	537	2436	1638
1967	545	2472	1701
1968	556	2560	1735
1969	558	2597	1771
1970	571	2592	1797
1971	585	2657	1894
1972	617	2801	2010
1973	642	2892	2117
1974	636	2842	2201
1975	632	2947	2265
1976	630	2983	2376

Sources: UK: *Economic Trends,* Central Statistical Office 1955–1979, HMSO, London.
US: *Statistical Abstract of the US,* 1957–1976, Government Printing Office.
Canada: *Canada Yearbook,* 1955–1976, Ministry of Supply and Services.

where y is income elasticity of food demand function and $\hat{p}$ is the pure price elasticity which is obtained by eliminating the income effect from the total price elasticity, p.

In order to find the required elasticity parameters in AIII.7 the food demand functions are specified as:

$$\ln D = \ln\beta + y\ln Y + p\ln P + tT \qquad \text{(AIII.8)}$$

for the UK and

$$\ln D = \ln\alpha + \mathrm{y}\ln Y + p\ln P \qquad \text{(AIII.9)}$$

for the US and Canada. D is an adult equivalent of per capita food demand, Y is real income measured in the same adult equivalent unit and P is the relative price of food in relation to non-food. The series for regression equations (AIII.8) and (AIII.9) are shown in Table AIII.3. The results are:

Country	y	p
UK	0.39	−0.54
US	0.51	−0.37
Canada	0.50	−0.41

Then, in order to estimate the pure elasticity, $\hat{p}$, the following procedure, supported by Stone (1954) and Fellner (1967), is employed. That is:

$$\hat{p} = y - (b)(p) \qquad \text{(AIII.10)}$$

where b is the share of food in consumer's budget (a proportion). Equation (AIII.10) is also the standard Slutsky equation expressed in terms of elasticities. The budget share of food, b, in the UK and US is 20 per cent and it is 17 per cent in Canada. Substituting these together with other estimates in (AIII.10) we get:

UK $\hat{p} = -0.47$
US $\hat{p} = -0.27$
Canada $\hat{p} = -0.32$

Then by using the equation (AIII.7) we get:

Country	*Elasticity of marginal utility of consumption,* e
UK	−0.70
US	−1.89
Canada	−1.56

Substituting all these estimates in equation (AIII.5) we get:

Country	*Social time preference rate* (%)
UK	2.6
US	5.3
Canada	5.4

Table AIII.3: Series for regression equations (AIII.8) and (AIII.9)

Year	Food demand, *D, per capita* adult			Income, *Y, per capita* adult			Relative food price, *P*		
T	UK (1970 £)	US (1967 $)	Canada (1961 $)	UK (1970 £)	US (1967 $)	Canada (1961 $)	UK (1970 = 100)	US (1967 = 100)	Canada (1961 = 100)
1954	121	504	285	481	2144	1512	101	104	101
1955	123	511	301	498	2271	1581	110	102	101
1956	130	483	315	501	2343	1650	110	101	101
1957	131	480	315	509	2351	1652	108	101	102
1958	131	469	315	521	2316	1659	107	103	104
1959	134	471	323	541	2420	1704	108	100	99
1960	136	533	327	559	2436	1708	105	99	99
1961	137	525	319	568	2428	1699	102	99	100
1962	137	529	320	575	2501	1745	97	99	101
1963	137	527	318	597	2561	1893	101	99	104
1964	138	535	325	613	2656	1850	101	99	105
1965	136	554	328	617	2774	1911	99	100	106
1966	138	557	321	628	2880	1957	99	102	110
1967	138	555	328	638	2917	2022	99	100	108
1968	137	560	324	650	3007	2054	98	99	106
1969	135	551	328	652	3041	2086	99	99	107
1970	134	560	336	667	3049	2108	100	98	106
1971	132	558	353	682	3095	2214	102	97	104
1972	128	588	359	718	3255	2343	104	98	109
1973	126	572	360	742	3352	2460	112	108	111
1974	125	552	355	738	3282	2552	115	113	125
1975	121	571	363	731	3283	2619	116	112	124
1976	118	590	385	731	3435	2743	121	110	118

Source: Kula (1984a, 1987b).

Appendix IV: Recent developments in forestry policy in the United Kingdom

Since the completion of this book some new developments have taken place in the United Kingdom.

These are as follows:

Proposed Farm and Woodland Scheme

Towards the end of 1987, a bill named the *Farm Land and Rural Development Bill* received its second reading in the House of Lords. Following this, the Minister of Agriculture, Fisheries and Food, the Right Honorable John MacGregor MP, announced the details of the proposed scheme in a press release.

The aims of the scheme are:

1. to divert land from agricultural production and thereby assist in the reduction of agricultural surpluses;
2. to enhance the landscape, to create new wildlife habitats, to encourage recreational use, including sport, and to expand tourist interests;
3. to contribute to supporting farm income and Natural employment;
4. to encourage greater interests in timber production from farms and, on a larger scale, to contribute to the United Kingdom's timber requirements.

Under this scheme annual payments would be made to farmers who plant trees on land previously in agricultural production. The scheme concentrates on arable land and improved grassland. That is to say that annual payments would be made available to farmers who plant trees on land previously in agricultural production. The first payment would be made one year after planting, and the payments would continue for a fixed period which would go someway towards bridging the gap between planting and the time when revenue normally arrives, for example from thinning. Annual payments would be in addition to, and dependent on, the payment of forestry authorities (Forestry Commission in Britain and the Forest Service in Northern Ireland) planting grants which are already available to farmers towards the cost of establishing woodlands. Their grants are payable at a higher rate for pure broadleaved woodland and for the broadleaves content in mixed woodland.

It was decided that the rate of annual grant should vary according to land quality, in order to reflect the fact that poorer land generally yields a lower income per hectare and that lesser incentives are needed to take it out of agricultural production.

The rates of annual payment under the proposed Farm Woodland Scheme are as follows:

Area	Grant per hectare per year
Outside the severely disadvantaged area, SDA, and disadvantaged area, DA, (mainly lowland)	£190
Disadvantaged area, DA	£150
Severely disadvantaged area, SDA	£100

It should be mentioned that there is not a common system of land classification in the United Kingdom. In England and Wales the agricultural land classification maps do not normally provide a sufficiently detailed assessment of grades for areas of less than 80 hectares and in Scotland also there are some areas which are not mapped in sufficient detail. To avoid anomalies, therefore, it would often be necessary to survey each site to determine the grade of land.

These payments would be made for 20 years for conifers, 30 years for woodland containing over 50 per cent broadleaves and 40 years in the case of woodland that is exclusively oak and beech.

The maximum area that will qualify for grant will be 12000 hectares a year, with an overriding limit of 36000 hectares during the first 3 years of the scheme. This is roughly half the recent annual rate of new planting under the traditional forestry programme in the whole of the United Kingdom. The government also intends to give some special help to the less favoured areas of the United Kingdom without upsetting the new thrust of the scheme. A total of 3000 hectares over the first 3 years, with the 36000 hectare limit, will be reserved for marginal land, including rough grazing, in less favoured areas. This would attract annual payments of £30 per hectare in view of the lower productive potential of such land. In addition, in Northern

Ireland, the minimum area to be planted under the scheme would be 1 hectare, rather than 3 hectares, to allow for the smaller average farm size.

The scheme would not apply to existing woodland areas. Farmers, however, continue to be encouraged to improve their management techniques and to rehabilitate derelict woodlands wherever possible. The forestry authorities planting grants would continue to be available for woodland regeneration where appropriate. In view of the persisting agricultural surpluses, the government believes that it is more desirable that the owners of derelict woodlands should improve their timber production capacity, rather than clear fell for conversion to agriculture. Planting of land which is not in productive agriculture would be excluded from the scheme. Applicants would be required to submit a map of the holding giving details of the productive agricultural areas to be planted and a declaration concerning previous agricultural production on these areas. Planting which would not form part of the applicant's normal farming operations, which involves the whole holding, christmas trees and trees harvested as saplings for nursery sale would also be excluded.

The Farm Woodland Scheme will come into operation in the autumn of 1988 following the passage of the Bill and the necessary secondary legislation. It is hoped that the scheme will encourage farmers to plant as many as thirty million trees per year throughout the United Kingdom.

Changes in taxation and planting grants

In his annual budget speech 15 March 1988, the Chancellor of Exchequer, the Right Honorable Nigel Lawson MP, stated that the tax system should recognise the special characteristics of forestry, where they may be anything up to a hundred years between the cost of planting and the income from selling the felled timber. However, he emphasized that the current system cannot be justified as it enables top-rate tax payers in particular to shelter other income from tax, by setting it against expenditure on forestry, while the proceeds from any eventual sale are effectively tax-free.

The Chancellor announced that as from 15 March 1988, and subject to transitional provisions, expenditure on commercial woodland will no longer be allowed as a deduction for income tax and corporation tax. But equally, receipts from the sale of trees or felled timber will no longer be liable to tax.

In parallel, there should be increases in planting grants. The net effect of these changes will be to end an unacceptable form of tax shelter; to simplify the tax system, and abolish the archaic schedule B in its entirety; and to enable the Government to secure its forestry objectives with proper regard for the environment, including the better balance between broad-leaved and coniferous trees.

On 22 March 1988 a new woodland grant scheme to provide government support for private forestry planting was announced by the Right Honorable Malcolm Rifkind QC MP, the Secretary of State for Scotland. In this, the Forestry Commission's Forestry Grant Scheme and Broadleaved Woodland Grant Scheme were declared closed to new applications from 15 March 1988, and will be replaced on 5 April 1988 by a single grant scheme, to be known as the Woodland Grant Scheme. This scheme will apply to the establishment and restocking of broadleaved, conifer and mixed woodlands, whether by planting or by natural regeneration, and to the rehabilitation of neglected woodland under 20 years of age. It will also cover planting land under the Farm Woodland Scheme.

The objectives of the new grant scheme are:

1. to encourage timber production;
2. to provide jobs in and increase the economic potential of rural areas with declining agricultural employment and to provide alternative sources of economic activity;
3. to provide an alternative to agricultural production and thereby assist in the reduction of agricultural surpluses;
4. to enhance the landscape, to create new habitats and to provide for recreation and sporting uses in the longer term;
5. to encourage the conservation and regeneration of existing forests and woodlands.

The new rates of grants for the whole of United Kingdom i.e. Great Britain and Northern Ireland, are:

Area approved for planting or regeneration, hectares	Rates of grant Conifers, per hectare	Broadleaves, per hectare
0.25–0.9	£1005	£1575
1.0 –2.9	£885	£1375
3.0–9.9	£795	£1175
10.0–and over	£615	£975

The rates of grant for conifer planting done under the Farm Woodland Scheme will remain unchanged; broadleaved planting under that scheme will, however, be eligible for the new broadleaved grants. These grants will be paid in three instalments over 10 years.

Apart from the increased rates of grant and special supplement for planting on better land, the scheme will have a number of important features. These are: the simplicity as the scheme covers all forms of planting, regeneration and rehabilitation of woodlands; a substantial differential in favour of broadleaves; all broadleaved planting, whether it be on its own or in mixture, will attract the same rates of grant; the broadleaved rate of grant will also apply to the planting and natural regeneration of native pinewoods in specified areas of Scotland; environmental objectives and provisions that will apply to all types of woodland; all types of natural regeneration will now be eligible for the first instalment of grant-aid at the time when the preparatory work is carried out; grants for the rehabilitation of derelict woodlands under 20 years of age will now include conifer and mixed woodlands; for conifers as well as for broadleaves, the second and third instalments will attract the rates of grant applicable when they fall due; grant bands will be determined by the total area approved for planting or regeneration under the Forestry Grant Scheme, by the size of the woodland of which such planting or regeneration might form a part.

Bibliography

Ahmad, I. and Laarman, J.G. (1978) Technologies for basic needs: the case of Philippine forestry, *International Labour Review, 77*, 491-499

Alfred, A.M. (1968) The correct yardstick for state investment, *District Bank Review*, No. 166, 21-32

Almon, S. (1965) The distributed lag between capital appropriations and expenditures, *Econometrica, 33*, 78-96

Annual Abstract of Statistics (1980) HMSO, London

Arrow, K.J. (1966) Discounting and public investment criteria. In *Water research*, eds. A.V. Kneese and S.C. Smith, Johns Hopkins University press, Baltimore, Md, pp. 13-32

Arrow, K.J. and Kurz, M. (1970) *Public investment, the rate of return and optimal fiscal policy*, Johns Hopkins University Press, Baltimore, Md

Boulding, K.E. (1935) The theory of a single investment, *Quarterly Journal of Economics, 49*, 475-494

Broussalian, V.L. (1966) Evaluation of non-marketable investments, *Research Contribution 9*, Centre for Naval Analysis, Arlington, Va

Broussalian, V.L. (1971) Discounting and evaluation of public investments, *Applied Economics, 3*, 1-10

Clawson, M. (1976) *Economics of natural forest management*, John Hopkins, Baltimore, Md

Clawson, M. (1979) Forests in the long sweep of American history. *Science, 204*, 1168-1174

Cmnd 1337 (1961) *The financial and economic obligations of nationalised industries*, HMSO, London

Cmnd 3437 (1967) *Nationalised industries: a review of economic and financial objectives*, HMSO, London

Cmnd 7131 (1978) *Nationalised industries*, HMSO, London

Cooper, J.P. (1972) Two approaches to polynomial distributed lags estimation: an expository note and comment, *American Statistician*, June, 32-35

Dasgupta, A.K. and Pearce D.W. (1972) *Cost benefit analysis: theory and practice*, Macmillan Student Editions, London

Eckstein, O. (1957) Investment criteria for economic development and

the theory of intertemporal welfare economics, *Quarterly Journal of Economics*, *71*, 56-83
Eckstein, O. (1961) A survey of theory of public expenditure. In *Public finances: needs, sources and utilisation*, ed. J. Buchanan, Princeton University Press, Princeton, NJ, pp. 451-502
FAO (1960–1985a) *Production Yearbook*, FAO, Rome
FAO (1969–1985b) *Yearbook of Forest products*, FAO, Rome
Farm Management Standards (1986) In *Agriculture in Northern Ireland*, Vol. 60, No. 9, HMSO, Belfast
Farm Management Survey (1980) *Farm Management Survey 1978/79*, University of Exeter, Agricultural Economics Unit, Exeter
Faustman, M. (1849) Berechnung des Wertes welchen Waldboden sowie noch nicht haubare Holzbestänce für die Waldwirtschaft besitzen, *Allgemeine Forst und Jagd-Zeitung*, *25*, 441-455
Feldstein, M.S. (1964) The social time preference rate in cost benefit analysis, *Economic Journal*, *74*, 360-379
Feldstein, M.S. (1974) Financing in the evaluation of public expenditure. In *Public finance and stabilisation policy — essays in honour of Richard A. Musgrave*, eds. W.L. Smith and J.M. Culbertson, North Holland, Amsterdam, pp. 13-26
Feldstein, M.S. and Fleming, J.S. (1964) The problem of time stream evaluation present value versus internal rate of return rules, *Bulletin of the Oxford University Institute of Statistics*, *26*, 79-88
Fellner, W. (1967) (ed.) Operational utility: the theoretical background and measurement. In *Ten economic studies in the tradition of Irving Fisher*, John Wiley, New York, pp. 39-75
Filius, A.M. (1986) The forestry incentives programme in the Netherlands. In *11th seminar on multipurpose agriculture and forestry*, University of Podova, Motto di Livenza, Italy
Fisher, I. (1927) A statistical method for measuring marginal utility and justice of a progressive income tax. In *The economic essays contributed in honour of J. Bates Clarke*, Clarke Essays, Macmillan, London
Fisher, I. (1930) *The theory of interest*, Macmillan, London
Forest Service (1975–1986) *Annual Reports*, Department of Agriculture Publications, Belfast
Forestry and British Timber (1986) February and September issues, London
Forestry Commission (1963a) *Bulletin 34*, HMSO, London
Forestry Commission (1963b) *Bulletin 35*, HMSO, London
Forestry Commission (1963c) *Booklet 43*, HMSO, London
Forestry Commission (1971) *Booklet 34, Forest Management Tables (Metric)*, HMSO, London
Forestry Commission (1972) *Forest Record 80*, HMSO, London
Forestry Commission (1974) *Bulletin 48*, HMSO, London
Forestry Commission (1975a) *Booklet 40*, HMSO, London
Forestry Commission (1975b) *Booklet 42*, HMSO, London
Forestry Commission (1976) *Booklet 43*, HMSO, London
Forestry Commission (1977a) *Wood production outlook in Britain*, HMSO, London

Forestry Commission (1977b) *Leaflet 64*, HMSO, London
Forestry Commission (1978) *Leaflet 70*, HMSO, London
Forestry Commission (1979) *Leaflet 72*, HMSO, London
Forestry Commission (1981) *Booklet 48*, HMSO, London
Forestry Commission (1980–1986) *Annual Reports*, HMSO, London
Forestry Review Group (1985) *Report to Minister for Fisheries and Forestry*, The Stationery Office, Dublin
Frisch, R. (1932) *The new methods of measuring marginal utility*, Verlag von J.C.B. Mahr, Tübingen
Goundrey, G.K. (1960) Forest management and the theory of capital, *Canadian Journal of Economics*, *26*, 439-451
Grayson, A.J. (1967) Species, growth rate and profitability, *Timber Grower*, January, 20-27
Grayson, A.J. (1974) Wood research and demand; statistical review, *Forestry Commission Forest Record*, *95*, HMSO, London
Hampson, S.F. (1972) Highland forestry: an evaluation, *Journal of Agricultural Economics*, *23*, 49-57
Hart, C.E. (1979) *British timber prices and forestry costings*, published by the author, Chennies, Coleford, Forest of Dean, Gloucester GL16 8DT
Hart, C.E. (1987) *The taxation of woodlands*, published by the author, Chennies, Coleford, Forest of Dean, Gloucester GL16 8DT
Hartwick, J.M. and Olewiler, N.D. (1986) *The economics of natural resource use*, Harper & Row, New York
Helliwell, D.R. (1974) Discount rates in land use planning, *Forestry*, *47*, 147-152
Helliwell, D.R. (1975) Discount rates and environmental conservation, *Environmental Conservation*, *2*, 199-201
Hiley, W.E. (1930) *Economics of forestry*, Clarendon Press, London
Hiley, W.E. (1967) *Woodland management*, Faber & Faber, London
HMSO (1972a) *Forestry in Great Britain: an interdepartmental cost benefit study*, HMSO, London
HMSO (1972b) *Forestry policy*, HMSO, London
Hotelling, H. (1925) A general mathematical theory of depreciation, *Journal of the American Statistical Association*, *20*, 340-353
Hirshleifer, J., De Haven, J.C. and Milliman, J.W. (1960) *Water supply: economics, technology and policy*, University of Chicago Press, Chicago, Ill.
Joint Economic Committee, US Congress (1968) *Economic analysis of public investment decisions: interest rate policy and discounting analysis*, US Government Printing Office, Washington, DC
Journal of Agricultural Statistics for Northern Ireland (1946–1984) Department of Agriculture, Belfast
Kay, J.A. (1972), Social discount rates, *Journal of Public Economics*, *1*, 359-378
Krutilla, J. and Eckstein, O. (1958) *Multipurpose river development*, Johns Hopkins University Press, Baltimore, Md
Kuhn, T.E. (1962) *Public enterprise economics and transport problems*, University of California Press, Los Angeles, CA
Kula, E. (1981) Future generations and discounting rules in public

sector investment appraisal, *Environment and Planning A, 13*, 899-910

Kula, E. (1984a) Derivation of social time preference rates for the United States and Canada, *Quarterly Journal of Economics, 19*, 873-883

Kula, E. (1984b) Discount factors for public sector investment projects by using the sum of discounted consumption flows method, *Environment and Planning A, 16*, 689-694

Kula, E. (1984c) Justice and efficiency with the sum of discounted consumption flows method, *Environment and Planning A, 16*, 835-838

Kula, E. (1985) Derivation of social time preference rate for the United Kingdom, *Environment and Planning A, 17*, 199-212

Kula, E. (1986a) The social time preference rate for Portugal, *Economia, 9*, 447-466

Kula, E. (1986b) A rate of return analysis for public forestry in Ulster, *Resources Policy, 12*, 346-350

Kula, E. (1986c) A cost benefit analysis of Ulster forestry, *The Royal Bank of Scotland Review*, No. 152, 32-39

Kula, E. (1986d) The analysis of social interest rate in Trinidad and Tobago, *Journal of Development Studies, 22*, 731-739

Kula, E. (1986e) The developing framework for the economic evaluation of forestry projects in the United Kingdom, *Journal of Agricultural Economics, 37*, 365-377

Kula, E. (1987a) The developing framework for the economic evaluation of forestry in the United Kingdom — a reply, *Journal of Agricultural Economics, 38*, 501-504

Kula, E (1987b) The social interest rate for public sector project appraisal in the U.K., the U.S. and Canada, *Project Appraisal, 2*, 169-175

Kula, E. and McKillop, D. (1988) A planting function for private afforestation in Northern Ireland, *Journal of Agricultural Economics, 39*, 133-141

Kula, E. (1988) The modified discount factors for project appraisal in the public sector, *Project appraisal, 3*, number 3

Kumar, R. (1982) World tropical wood trade, economic overview, *Resources Policy, 8*, 177-192

Land Use Study Group (1966) *Forestry, agriculture and multiple use of rural land*, HMSO, London

Lanly, J.P. (1979) Present and future of natural forests and plantation areas in the tropics, *Unasylva, 31*, 12-20

MacGregor, J.J. (1953) The preparation of index numbers for forestry products, *Forestry, 32*

McEwan, J. (1981) *Who owns Scotland?* Polygon Books, London

McKillop, D. and Kula, E. (1987) The importance of lags in determining the parameters of a planting function for forestry in Northern Ireland, *Forestry, 60*, 229-237

Marglin, S. (1963a) The opportunity cost of public investment, *Quarterly Journal of Economics, 77*, 274-289

Marglin, S. (1963b) The social rate of discount and the optimal rate of

saving, *Quarterly Journal of Economics*, *77*, 95-111
Midland Bank (1980) *Forward Planning with the Midland Bank Group*, Midland Bank Agricultural Development and Advisory Service, London
Miller, R. (1981) *State forestry for the axe*, Institute of Economic Affairs, London
Ministry of Overseas Development (1977) *A Guide to Economic Appraisal of Projects in Developing Countries*, HMSO, London
Nash, C.A. (1973) Future generations and the social rate of discount, *Environment and Planning A*, *5*, 611-617
Northern Ireland Abstract of Statistics (1982–1984) HMSO, Belfast
Northern Ireland Digest of Statistics (1968–1981) HMSO, Belfast
Page, T. (1977) Equitable use of resource base, *Environment and Planning A*, *9*, 15-22
Pigou, A. (1929) *The economics of welfare*, Macmillan, London
Potter, N. and Christy, F.T. (1962) *Trends in natural resource commodities; statistics of prices, output, consumption, foreign trade and employment in the US, 1870–1977*, Johns Hopkins University Press, Baltimore, Md
Price, C. (1973) To the future — with indifference or concern?, *Journal of Agricultural Economics*, *24*, 383-398
Price, C. (1976) Blind alleys and open prospects in forestry economics, *Forestry*, *49*, 93-107
Price, C. (1978) *Landscape economics*, Macmillan, London
Price, C. (1981) Some economic aspects of marine management policies, the future and discount rate, *FAO Fisheries Series No. 5*, pp. 57-65, FAO, Rome
Price, C. (1984a) Project appraisal and planning for overdeveloped countries — the costing of nonrenewable resources, *Environmental Management*, *8*, 221-232
Price, C. (1984b) The sum of discounted consumption flows method; equity with efficiency?, *Environment and Planning A*, *16*, 829-837
Price, C. (1987) The developing framework for the economic evaluation of forestry in the United Kingdom — a comment. *Journal of Agricultural Economics*, *38*, 497-500
Rawls, J. (1972) *A theory of justice*, Clarendon Press, Oxford
Reed, F.L. (1979) Optimal escapement levels in stochastic and deterministic harvesting models, *Journal of Environmental Economics and Management*, *6*, 350-366
Samuelson, P.A. (1976) Economics of forestry in an evolving society, *Economic Inquiry*, *14*, 466-492
Simonis, U.E. (1984) Environmental crisis — the missing dimension in the North–South dialogue, *Economics*, *30*, 48-64
Soper, C.S. (1959) The marginal efficiency of capital; a further note, *Economic Journal*, *69*
Steiner, P.A. (1959) Choosing among the alternative investments in the water resources field, *American Economic Review*, *49*, 893-916
Stone, R. (1954) *Measurement of consumer expenditure and behaviour in the United Kingdom, 1920–1928*, Cambridge University Press, Cambridge

Syson, W.C. (1985) Forestry financing — Scottish bank plays major role, *Forestry and British Timber*, September
Thompson, A.E. (1971) The Forestry Commission: a reappraisal of its functions, *Three Banks Review*, September, 30-44
Tompkins, S.C. (1986) *The theft of the hills*, Ramblers Association, London
Von Thunen, J.H. (1926) *Isolated state*, English edition, ed. P. Hall, Pergamon Press, London
Walker, K.R. (1958) Competition for hill land between the agriculture, industry and Forestry Commission, unpublished PhD thesis, Oxford University
Wolfe Report (1973) *Some considerations regarding forestry policy in Great Britain*, Forestry Committee for Great Britain, Edinburgh
Wright, J.F. (1963) Notes on the marginal efficiency of capital, *Oxford Economic Papers, 15*, 124-129

Author Index

Subject Index